Help One Another

Volume 2

Cook Book

PUBLISHED BY THE
DAKOTA FARMER
AND NORTH PLAINS PRESS

EDITED BY
Mary Bennett
and Jean Laustsen

Library of Congress Catalog Card Number 73-77360
ISBN Number 0-87970-129-3
Manufactured in the United States of America

North Plains Press
Aberdeen, South Dakota

INTRODUCTION

It is with great pride and eagerness that THE DAKOTA FARMER magazine gives you its collection of authentic prairie recipes, straight from the family kitchens of North and South Dakota. Carefully selected from the thousands and thousands of recipes sent to our "Help One Another" column by subscribers to THE DAKOTA FARMER, the recipes included in this cookbook are some of the finest Dakota has to offer. And no wonder . . . they are all tried and true family favorites, many passed down from generation to generation of fine Dakota cooks.

Dakota cooking is seasoned with paradox. While it is delightfully American on one hand, it is deliciously European on the other. Less than a century ago Dakota was not only an American frontier, it was a European frontier as well — a vast open prairie beckoning to hundreds of thousands of "Old World" immigrants.

It was these hard-working, industrious settlers, working hand-in-hand with American pioneers from the East, who broke the prairie sod and planted their crops. Food was indeed a precious commodity, and cooking it to perfection was a way of life. Some of the most imaginative cooks were farmers' wives who shared a love of the land and its produce, creating meals which were at once healthy, rugged, hearty, and delicious.

It was from these proud and humble beginnings that Dakota cooking got its start. Here, in the pages that follow, are collected some of Dakotas' most popular "heirloom" recipes, as well as a vast collection of new and inventive cooking from the kitchens of Dakota today.

If the smell of home-baked bread makes your taste-buds tingle; if the sight of a freshly-baked pie sets your stomach agrowl; if the thought of piping hot lefse fills you with eager anticipation, then you have hours and hours of happy baking and good eating ahead.

But don't thank us. Thank the many talented and creative cooks on the plains of North and South Dakota.

— Jean Laustsen and Mary Bennett,
Editors

LAWRENCE WELK'S FAVORITE RECIPE

CREAMED CHICKEN

1 medium fryer, cut-up
  salted and peppered
1 tbsp. butter
1 small onion
½ pint cream

Melt butter in frying pan and add chicken. Keep stirring to sear, but not brown. Then add onion, finely chopped; and let simmer for about 1 minute. Add 2 tablespoons water. Cover and simmer slowly for 30 minutes. If more water is needed during cooking, add just enough to keep meat stewing. After 30 minutes remove cover and add cream. Be sure not more than a tablespoon of water is left in pan before adding the cream. Stir meat in cream several times during last 5 minutes. Simmer for 5 minutes more and remove from heat. Serve immediately.

A Foreword by Lawrence Welk

I'm honored and flattered to be asked to write a foreword for a cook book, although I must confess my qualifications are somewhat questionable. You see, when it comes to food, my role is strictly that of a consumer.

Those of you familiar with our Television Show know that simplicity and wholesomeness are two essential qualities we always strive for. I guess it's only natural that my tastes in food would follow similar lines. "Gourmet dining" was not something you learned on a farm in the Dakotas. To be perfectly honest, it's something I've never learned.

In the early days of our band, (the late 20's and 30's) I guess we played every community of any size in South Dakota, and we were always received with the very warmest hospitality and friendliness. One of the "fringe benefits" of our job in those days was an invitation to share a home-cooked meal at the home of one of the kind hearted farmers of the area. In some cases this constituted our entire salary for our work. I can truly say, I never felt underpaid!

Unless Dakota home cooking has deteriorated greatly in recent years, (which I seriously doubt) I can give it my wholehearted endorsement, with the devout hope it never becomes a lost art.

DEDICATION

To the true authors of this book, the many thousands of Dakotans who have shared their favorite recipes with THE DAKOTA FARMER.

Table of Contents

DAKOTA HERITAGE

In the Dakotas some of the best all-American cooking is straight from the recipe files of our German and Scandinavian ancestors. The smell of fresh kuchen, knepfla or lefse is as common and delightful an experience in Dakota today as it was years ago. Here, in the pages that follow, are some of the most popular Scandinavian and German heirloom recipes, as well as some favorite recipes from another important Dakota ancestor, the American Indian.

Czechoslovakian
German

American Indian
Scandinavian

Czech/**DAKOTA HERITAGE**

KOLACE

- 2 ozs. fresh yeast or 2 pkgs. dry yeast
- ½ c sugar
- ¼ c lukewarm water
- 6 c flour
- 2 tsps. salt
- 3 c milk, scalded and cooled
- ½ c evaporated milk
- 2 eggs
- ½ c shortening, melted
- filling

Dissolve yeast and 3 tablespoons sugar in lukewarm water. Combine flour, remaining sugar and salt. Combine milk and eggs; beat well. Add shortening; beat again; add to flour mixture and mix well. Knead until dough is smooth and elastic, sprinkling with flour a little at a time. Brush with melted shortening; cover. Put in warm place to rise until double in bulk. Shape into balls about the size of a walnut. Place on well-greased pan, spacing about 15 balls on 10x15-inch pan. Brush tops with melted shortening; let rise in warm place until light. Make depression in each ball; put in filling. Return to warm place to finish rising. Bake at 400-425 degrees for 12-15 minutes or until golden brown. Brush with melted shortening or sugar-water glaze. Serve.

Fruit Filling: Cook 2 pounds fruit (dried prunes, apricots, peaches, etc.) in water until tender. Drain; cool; pit. Grind in food chopper; add 1 cup sugar, 1 teaspoon vanilla and a dash of cinnamon.

Cottage Cheese Filling: Press all liquid out of 1 pound creamed cottage cheese until it is a thick paste. Add 1 tablespoon butter and ½ cup sugar; cream. Add 1 egg yolk, a pinch of salt and ½ teaspoon vanilla (or lemon flavoring). Sprinkle lightly with cinnamon; mix well.

Poppy seed filling: To a 12-ounce can of poppy seed filling, add ½ cup prune filling; mix well.

Miss E. W., Hutchinson Co., S. D.

German/**DAKOTA HERITAGE**

FLEISH KIECHLA

- 3 c flour
- 1 tsp. salt
- 1½ c milk
- 1 c ground beef
- 1 c ground pork
- 1 small onion, chopped
- salt and pepper
- ½ c cold water

Roll dough of first three ingredients into thin circles. Mix remaining ingredients for filling. Spread a small amount on half of the circle, fold over, seal edges and deep fry.

Mrs. J. K., Hettinger Co., N. D.

KNEPFLA

4 c flour
2 eggs
1 tsp. salt
1 tbsp. salt
boiling water
butter and onions or 1 cup cream and 1 onion, diced

Cheese Filling: (Optional)
4 c cottage cheese
1 egg
salt
pepper
½ c sugar (optional)

Make stiff dough of flour, eggs and 1 teaspoon salt. Add 1 tablespoon salt to pan of boiling water. Cut small pieces of dough; drop into boiling water. Cook until done or until knepflas come to the top of water. Drain and fry in butter and onions or boil together diced onion and cream and pour over knepfla. Heat and serve.

Variation: This same dough may be used to make kase knepfla. Roll out small pieces of dough. Mix together cheese filling ingredients; put 1 tablespoon on each piece of dough. Fold over and press edges together. Drop one by one into large kettle of boiling salted water. When done fry in butter or boil cream and pour over them; heat.

Mrs. F. S., Grant Co., N. D.

SEIBEL SALAD

20 marshmallows, rinsed and diced
1 (No. 2) can pineapple chunks, drained
¼ lb. pasteurized process cheese, cubed
3 tbsps. flour
1 egg
pinch of salt
juice from can of pineapple

Combine marshmallows, pineapple chunks and cheese. Combine flour, egg, salt and pineapple juice; cook until thick. Add to marshmallow mixture; mix well.

Mrs. J. S., Ward Co., N. D.

PLUMA MOOS

1 lb. dried fruit
1 qt. water
1 c sugar

Place any combination of prunes, raisins, apricots, apples, etc., in kettle. Cover with water; simmer until fruit is tender. Add sugar and stir until sugar is dissolved. Chill and serve.

Variation: Use only 2 cups water on fruit. For sauce, mix 2 cups half-and-half, 3 tablespoons flour, ½ teaspoon cinnamon and ½ teaspoon salt. Heat to boiling; mix with fruit. Chill and serve.

Miss E. W., Hutchinson Co., S. D.

NEUFCHATEL CHEESE (GOAT MILK)

Heat 1 gallon goat milk to 72 degrees. Add 1 teaspoon soured milk and 1/10 rennet tablet dissolved in ½ cup cold water. Let set 18 hours. Pour curd into cloth bag; drain in clean cool place for 12-48 hours. When curd is firm, salt to taste.

Variation: You may add finely ground pimiento or peppers. In making pimiento Neufchatel cheese, add 7-10 drops cheese color to milk before adding rennet tablet. It gives a richer appearance.

E. C., Dewey Co., S. D.

APFELSTRUDEL

- 2½ c flour
- 1 tsp. salt
- 2 tbsps. shortening
- 2 eggs, slightly beaten
- ½ c warm water
- 5 c sliced apples
- 1 c brown sugar
- ½ c raisins
- ½ c chopped nuts
- 3 tbsps. butter, melted
- ½ tsp. cinnamon

Sift together flour and salt. Cut in shortening. Add eggs and warm water. Knead well until batter blisters. Cover with cloth; put in warm place for 20 minutes. Place white cloth on table; sprinkle with flour. Put dough on cloth; pull out very carefully with hands to thickness of tissue paper. Combine remaining ingredients; spread onto dough. Roll by using cloth. Bake at 450 degrees for 10 minutes. Reduce heat to 400 degrees and bake 20 minutes longer. Cool and slice.

Variation: Cottage cheese, raisins, cream and sugar; or pumpkin slices, raisins and sugar can also be used as filling.

Mrs. P. M., Stark Co., S. D.

GROSEN KLUMP

- 4-5 eggs, beaten
- ½ tsp. salt
- ¼ tsp. ground cardamom
- 1 tbsp. sugar
- 2 c milk
- 3½ c flour
- 2/3 tsp. baking powder
- 2-3 tbsps. butter, melted
- 1 c raisins

Mix together all ingredients except raisins. Place raisins in a cloth; add first mixture. Tie cloth; cook in boiling water for 1½ hours. Remove from cloth immediately; serve with melted butter.

Mrs. L. K., Minnehaha Co., S.D.

DAKOTA HERITAGE/German

HOT GERMAN POTATO SALAD

6 medium potatoes, boiled in jackets
6 slices bacon
¾ c chopped onion
2 tbsps. flour
½ tsp. sugar
1½ tsps. salt
½ tsp. celery seed
dash of pepper
¾ c water
1/3 c vinegar

Peel potatoes and slice thin. Fry bacon slowly in skillet; drain on paper. Saute onion in bacon fat until golden brown. Blend in flour, sugar, salt, celery seed and pepper. Cook over low heat, stirring until smooth and bubbly. Remove from heat; stir in water and vinegar. Heat to boil, stirring constantly. Boil 1 minute. Carefully stir in potatoes and crumble bacon bits. Remove from heat; cover until ready to serve.

Miss E. W., Hutchinson Co., S. D.

GRUN SCHAUBLE SUP (Green String Bean Soup)

Cook until tender ham bone or piece of ham. Add cubed potatoes, diced carrots, green beans and several sprigs of summer savory. When vegetables are well done, add ½ cup sour cream. Stir well and serve.

Miss E. W., Hutchinson Co., S. D.

BLANCHINDA

Dough:
6 c flour
1¼ c cold lard
2 tsps. salt
2 tsps. sugar
2 tsps. baking powder
1¼-1½ c ice cold water

Filling:
2 qts. cooked, drained and mashed pumpkin
1¼ c brown sugar
1 tbsp. cinnamon
1 tbsp. salt
¼ c chopped onion, sauted in 4 tbsps. butter
½ tsp. pepper (optional)

With a pastry blender, mix together dough ingredients in order given. Divide into 6 portions. Roll dough medium thin; cut into 6 pieces. Prepare filling by mixing together all filling ingredients. Place 1/3 cup or more of pumpkin mixture on each cut piece of dough. Moisten edges with water; pinch edges tightly. Place on cookie sheet; pierce each blachinda with sharp fork. Bake at 350 degrees for 30 minutes.

Mrs. R. B., Hutchinson Co., S. D.

KUCHEN

2 c warm milk
½ c sugar
½ c shortening
1 tsp. salt
1 pkg. dry yeast
2 eggs
flour

Filling:
2 c sweet or sour cream
2 eggs, beaten
½ c sugar
1½ tbsp. flour
1 tsp. vanilla

Crumbs:
1c sugar
1 c flour
½ c butter

Mix together dough ingredients, using enough flour to make a soft dough. Separate dough into 8 equal parts. Roll each piece out to fit pie tin. Let rise for about 20 minutes. Fill each with cream filling. Put any fruit toping on top (apple, peach, apricot, grape, prune, etc.). Sprinkle crumbs over top; sprinkle with sugar and cinnamon. Bake at 350 degrees until brown.

Filling: Heat cream. Combine eggs, sugar and flour; stir into cream and cook until thick. Add vanilla; cool slightly.

Crumbs: Cut together crumb ingredients.

Mrs. B. F., Jr., Jerauld Co., S. D.

Kuchen Variations

Cottage Cheese Kuchen: Use same dough and filling, but mix cottage cheese with filling. May add raisins or prunes or ½ teaspoon dried and ground orange peel. Sprinkle with paprika or cinnamon.

Holiday Kuchen: Use same dough, but make filling by mixing together 2 tablespoons butter, ¼ cup sugar, 2 teaspoons flour, 1½ tablespoons cream, chopped hazel nuts or pecans and maraschino cherries or coconut.

Sugar Kuchen: Roll dough to cover only bottom of pie pan; let rise. Spread with mixture of 1-2 tablespoons butter or sour cream, 6 tablespoons sugar and 1 tablespoon flour. Sprinkle with cinnamon and bake; or sprinkle dough with milk and sugar and bake.

SAUER KRAUT

1 qt. kraut
1 c water
¼ c sugar
2 tbsps. lard
2 tbsps. flour

Combine kraut, water and sugar; heat to boil; simmer for 5 minutes. Make sauce by heating lard until very hot; add flour; brown. Add sauce to kraut; mix well and serve.

Miss E. W., Hutchinson Co., S. D.

INDIAN FRIED BREAD

2 c milk, scalded
1 tbsp. honey
½ tsp. salt
2 eggs, beaten
2 tbsps. butter
1 pkg. dry yeast
¼ c warm water
6 c sifted flour

Combine milk, honey, salt and butter; add beaten eggs. Soften yeast in warm water. Add 3 cups flour to milk mixture; beat well. When mixture is cool, add yeast and remaining 3 cups flour; beat well. Grease top well; let rise until double in bulk. Roll out on floured surface. Cut in 3-inch strips; slash centers; fry in deep fat, turning once. Drain on absorbent paper. Serve with ham, bacon and eggs, honey and syrup.

Mrs. V. P., Brown Co., S. D.

BAKED HEART WITH BREAD DRESSING

1 beef heart cut-up
salt
Bread Dressing:
½ loaf bread, cubed
salt
pepper
sage
chopped onion

Cover beef heart with water; cook until tender. Remove from heat. Combine bread dressing ingredients; add broth from boiled heart. Mix heart with dressing and serve.

Winnebago Reservation

ROSEBERRY JAM

4 c roseberries
3 c boiling water
1 lemon, sliced
3 c sugar

Cook roseberries in water for 30 minutes; mash. Cook lemon in small amount of water for about 1 minute. Combine berries, lemon and sugar; cook until thick. Pour into hot boiled jars; seal.

Yankton Reservation

WASNA

1 c dried meat
1 c dried chokecherries
sugar
salt
hot bone grease or suet

Roast dried meat in oven until brittle. Pound it until soft and fine. Soak cherries in small amount of hot water. Pound cherries; mix with meat. Add sugar and salt to taste. Pour hot bone grease or suet over mixture; serve at once.

Pine Ridge Reservation

SWEDISH COOKIES

1 c butter
2 c sifted flour
½ c potato flour
1/3 c sugar

Blend together all ingredients as for pie crust. Roll into 2 rolls; dip in colored sugar or chopped nuts. Chill. Slice and bake at 370 degrees until done.

Mrs. S. N., Mountrail Co., N. D.

SPRITZ

¾ c butter
½ c sugar
1 egg yolk
¼ tsp. salt
½ tsp. orange extract
1 tsp. grated orange rind
2-2¼ c sifted flour

Cream together butter and sugar thoroughly; beat in egg yolk. Add extract and rind. Sift together salt and flour; add to above mixture. Put through cookie press onto ungreased cookie sheet. Bake at 425-450 degrees for 8-10 minutes.

Variation: For chocolate spritz, add 5 tbsps. cocoa and use the whole egg.

Mrs. D. G., Tollett Co., N. D.

RUSK

8 eggs, beaten
½ c sugar
1½ c lukewarm water
1 pkg. yeast
3 tsps. salt
½ c butter
½ c shortening
8-10 c flour

Mix together ingredients in order given. Let rise; knead down; let rise again. Divide dough into 7 equal parts; fill 7 orange juice cans, lined with waxed paper, about ½ full. Let rise until almost filled. Bake at 350 degrees for 25-30 minutes. Remove from cans; cool. Slice 1-inch thick; toast on both sides under broiler.

Mrs. E. J. M., Douglas Co., S. D.

LIVER PUDDING (Rospe Polsa)

¾ lb. veal liver
4 medium potatoes
1 small onion
1¼ c raisins
¼ c fat, melted
1 tsp. ginger
¼ tsp. ground mace

Simmer veal liver in enough water to cover for 10 minutes. Put meat through grinder with potatoes, onion and raisins. Add fat and seasonings; blend thoroughly. Place in buttered baking dish. Bake at 350 degrees for 25-35 minutes, or until browned.

Mrs. G. B., Mercer Co., N. D.

SWEDISH RYE BREAD

½ c warm water
2 pkgs. dry yeast
1½ c lukewarm milk
2 tbsps. sugar
1 tbsp. salt
2 tbsps. shortening
½ c molasses
3¼ c flour
3¼ c medium rye flour

Soak yeast in warm water for 5 minutes. Combine milk, sugar and salt in bowl; stir to dissolve. Beat in shortening, molasses, white flour and softened yeast. Add rye flour. Turn out on floured board. Knead until dough is smooth and elastic and no longer sticky. Let rise until double in bulk (about 1-1½ hours at 80-85 degrees). Punch down; let rise again until nearly double. Divide dough in 2 parts; shape into 2 round loaves. Let rise until double (1-1½ hours). Bake at 375 degrees for 30-35 minutes.

Mrs. H. K., Kingsbury Co., S. D.

BAKED LUTEFISK

Place fish on drain board; let water drain off for about 2 hours. Place fish in roaster, skin side down. Sprinkle with salt. Cover and bake at 400 degrees for 40 minutes.

Mrs. A. A., Marshall Co., S. D.

FLAT BREAD

2 c corn meal
4 c flour
2 c old-fashioned oatmeal
1¼ tsps. soda
3 tsps. salt
4 tbsps. sugar
1 qt. buttermilk
1 c sour cream
1 1/3 c shortening

Combine corn meal, flour, oatmeal, soda, salt and sugar; set aside. Mix together buttermilk, sour cream and shortening; boil. Pour over flour mixture; mix well. Let stand until cool. Add enough flour to roll. Roll out very thin. Bake on lefsa pan until done.

Mrs. A. T., Codington Co., S. D.

NORDLAND LEFSE

2 c milk, scalded
½ c lard
1 tbsp. salt
4 c flour

Combine milk, lard, salt and 2 cups flour; mix well over low heat. Remove from heat. Sift 2 cups flour on board; add warm dough; work in flour, kneading well and cooling. Make a small patty, about the size of a hamburger, with low sealed edges as you would for pastry. Flour board lightly and roll patty in all directions. Stretch dough until it is almost as large as lefse baker. Brown lightly on lefse baker, pinching out air bubbles; turn and brown other side. Serve with butter or butter, sugar and cinnamon.

Mrs. A. B., Traill Co., N. D.

BREADS & ROLLS

Fresh bread, hot from the oven and spread with rich butter or jam, outshines even the most elegant cake or pastry. Simple, nutritious and taste-tempting, breads have always held an important place on Dakota dinner tables. Today's farmwife carries on her pioneer grandmother's bread-baking tradition. The equipment and methods are modernized, the variety of yeast and quick breads larger, but plain or fancy, bread is still of number one importance in farm cookery.

Quick — Breads

- **Bisquits**
- **Coffee Cakes**
- **Doughnuts**
- **Muffins**
- **Pancakes & Waffles**

Yeast — Breads

- **Doughnuts**
- **Rolls & Buns**
- **Coffee Cakes**

YEAST STARTER

1 pkg. active dry yeast (or
  1 cake compressed yeast)
½ c warm water
2 c lukewarm water
2 tsps. sugar
1 tsp. salt
2 c flour

Soften yeast in ½ cup warm water. Add remaining ingredients; mix well. Cover; let stand three days at room temperature. Stir down each day. On fourth day starter is ready to use. To make bread use 1½ cups starter.

Note: To remaining starter add 1½ cups lukewarm water, ½ cup flour and 1 teaspoon sugar. Cover; store in refrigerator until next time you bake. Remove starter from refrigerator 3-12 hours before using.

Mrs. J. F. B., Roberts Co., S. D.

DARK BREAD

1 pkg. dry yeast
3¼ c lukewarm water
1 tsp. sugar
4 c flour
1/3 c lard, melted
¾ c molasses
½ c sugar
2 tsps. salt
3½ c fine ground whole
  wheat flour

Dissolve yeast in ¼ cup lukewarm water. Add 1 teaspoon sugar. Combine remaining 3 cups lukewarm water and 2 cups flour. Add yeast; allow to rise about 20 minutes. Add lard, molasses, ½ cup sugar and salt. Stir in whole wheat flour. Knead in 2 cups white flour. Allow to rise until double in bulk. Punch down; allow to rise again until almost double in bulk. Shape into 2 loaves. Allow to rise until almost double. Bake at 350 degrees for 50-60 minutes.

Mrs. J. H., Walsh Co., N. D.

NEVER-FAIL BREAD

3 pkgs. fast-rising yeast
  or 2 ozs. fresh yeast
5½ c warm water
½ c sugar
4 tsps. salt
15 c flour
1/3 c lard or shortening

Dissolve yeast in ½ cup warm water. Set aside for 5-10 minutes. Mix together 5 cups warm water, sugar and salt; stir until dissolved; add yeast-water mixture. Add 8 cups flour; mix well. Add shortening; beat well. Add remaining 7 cups flour; beat well. Knead until smooth. Grease top of bread mixture. Cover; let bread rise until double in bulk. Knead; let rise until double in bulk. Remove dough from bowl; roll with rolling pin to distribute air bubbles evenly. Shape into loavesand buns; let rise until double in bulk. Bake at 375 degrees for 50 minutes for bread, 20-25 minutes for buns. Makes 6 loaves of bread, or 4 loaves and 2 pie pans of buns.

Mrs. J. K., Stutsman Co., N.D.

YEAST BREADS

FRENCH BREAD

- 2 pkgs. dry yeast
- ½ c warm water
- 7-7½ c sifted flour
- 1 tbsp. salt.
- 2 c lukewarm water
- 1 egg white
- 1 tbsp. water

Soften yeast in ½ cup water. Add salt to 2 cups lukewarm water; beat in 2 cups sifted flour. Blend in softened yeast; stir in 4-4½ cups sifted flour, or enough to make soft dough. Turn out on lightly floured surface. Cover; let rest 10 minutes. Knead until smooth and elastic (5-8 minutes) working in last cup of flour. Place in lightly greased bowl, turning dough once. Cover; let rise until double in bulk (1½ hours). Punch down; let rise until double again (about 1 hour). Punch down; turn out on lightly floured surface; divide into 2 portions. Cover; let rest 10 minutes. Roll each portion into 15x12-inch rectangles. Roll tightly, beginning at long side, sealing well. Taper ends. Place each loaf diagonally, seam side down, on greased baking sheet that has been sprinkled with corn meal. With sharp knife gash tops diagonally every 2½ inches, ⅛-¼-inch deep. Beat egg white until foamy; add 1 tablespoon water. Brush mixture over tops and sides of loaves. Cover with damp cloth, but don't allow to touch loaves. Let double (1½ hours). Bake at 375 degrees for about 20 minutes, or until light brown. Brush again with egg white mixture. Bake about 20 minutes longer.

Mrs. C. S., Hettinger Co., N. D.

WHITE BREAD

- 2 pkgs. active dry yeast
- 1 c warm water
- 6 tbsps. sugar
- 2 tbsps. salt
- ¼ c soft shortening
- 3½ c lukewarm milk, water or potato water
- 14-14½ c flour

In a large bowl, dissolve yeast in water. Add sugar, salt, shortening and milk, water or potato water. (When using milk, scald and cool to lukewarm.) Mix; add about half the flour; beat until smooth. Add remaining flour until dough clings to bowl. Turn onto lightly floured board; cover and let rest for 10 minutes. Knead until smooth and blistered (about 10 minutes). Place in greased bowl, bringing greased side up. Cover with cloth; let rise in warm place about 1-2 hours until double in bulk. Punch down. Cover; let rise until almost double, about 45 minutes. Divide into 4 parts. Round up and let rest about 10 minutes. Shape into loaves. Place in greased loaf pans, bringing greased side up. Cover with cloth; let rise until sides reach top of pan and top is rounded (50-60 minutes). Place loaves in center of oven, not touching each other or sides of oven. Bake at 425 degrees for 25-30 minutes, or until deep golden brown.

Note: For a crisper crust, do not grease or cover after baking. For tender crust, brush with shortening and cover with towel for a few minutes. For a glazed crust, brush with mixture of 1 egg yolk and 2 tablespoons water before baking.

Mrs. W. L., Pennington Co., S. D.

PUMPERNICKEL BREAD

- 1½ c warm water
- 2 pkgs. active dry yeast
- ½ c molasses
- 4 tsps. salt
- 1-3 tbsps. caraway seed
- 2¾ c sifted rye flour
- 3½-4 c sifted white flour

Dissolve yeast in warm water. Stir in molasses, salt and caraway seed. Mix in rye flour with a spoon until smooth. Add white flour; mix with hands to blend completely. Turn onto a lightly floured board; knead and let rise as for white bread. Shape into 2 slightly flattened loaves. Place on opposite corners of greased baking sheet; sprinkle with corn meal, if desired. Cover with damp cloth. Let rise 1 hour. Bake at 375 degrees for 30-35 minutes.

Mrs. C. S., Codington Co., S. D.

RYE BREAD

- 1 c graham flour
- ¼ c brown sugar
- 3 c rye flour
- 1 tbsp. salt
- hot water
- 1 cake yeast
- 1 c lukewarm water
- wheat flour

Mix together graham flour, brown sugar, rye flour and salt. Beat in enough hot water to form a stiff batter. Cover; let stand until mixture is lukewarm. Dissolve yeast in lukewarm water; add to flour mixture. Add enough wheat flour to mold. Let dough stand until double in bulk. When light, shape into loaves. Bake at 375 degrees for 1¼ hours.

Mrs. J. N., Potter Co., S. D.

YULEKAGE

- 1 cake yeast, dry or compressed
- ¼ c lukewarm water
- 1 c dried currants
- 1 c raisins
- 2/3 c almonds
- 2 c milk, scalded
- 1/3 c lard
- ¼ c sugar
- 2 tsps. salt
- 7 c flour
- 3 eggs, beaten
- ½ tsp. crushed cardamom seed
- 1 c finely cut candied citron

Dissolve yeast in lukewarm water. Prepare dried currants and raisins. (Raisins may be cut if desired.) Blanch almonds; spread out to dry; cut in pieces. Combine scalded milk, lard, sugar and salt. Cool to lukewarm. Add 2 cups flour; mix well. Add yeast mixture. Add eggs and cardamom seed. Mix together 1 cup flour, raisins, currants, almonds and candied citron; add to yeast mixture. Blend thoroughly. Add 4 cups flour or enough to make dough soft. Turn out on floured board; knead until smooth. Shape into ball. Place in greased bowl; put in warm place. Let rise until double in bulk (about 2 hours). Punch down. Divide dough into 3 portions; shape each portion into a ball. Cover and let stand 10 minutes. Shape each ball into loaf; place in 3 tins. Let rise until double in bulk. Bake at 350 degrees for 1 hour.

Mrs. E. J., Walsh Co., N. D.

YEAST BREADS/Doughnuts

BAKED DOUGHNUTS

1 c milk, scalded
¾ c shortening
½ c sugar
1 tsp. salt
1 cake or pkg. yeast
¼ c warm potato water
1 c mashed potatoes
2 eggs, beaten
4½ c flour

Combine milk, shortening, sugar and salt. Cool to lukewarm. Dissolve yeast in warm potato water. Add mashed potatoes and dissolved yeast to milk mixture. Blend in eggs and flour. Cover; let rise until light. Place dough on floured board; knead. Roll dough to about ½-inch thickness; cut with doughnut cutter. Place doughnuts on buttered cookie sheet; let rise. Bake at 425 degrees until brown. Brush with butter; sprinkle with sugar.

Mrs. A. R., Morton Co., N. D.

ICED DOUGHNUTS

2 cakes yeast
1 c warm water
¾ c milk, scalded
1 tsp. salt
¼ c sugar
1/3 c shortening, melted
4½ c flour
2 eggs, well-beaten

Icing:
1 lb. confectioners sugar
1 tbsp. cornstarch
1 tsp. vanilla
½ c butter, melted
hot water

Soften yeast in warm water for 5 minutes. Pour scalded milk over salt, sugar and melted shortening. Stir until cooled. Add 1 cup flour; beat until smooth. Beat in eggs, yeast and 3½ cups flour. Mix to soft dough. Turn onto lightly floured board. Knead until dough is smooth. Place in lightly-greased bowl. Grease top of dough. Cover; allow to rise in warm place until doubled in bulk (1-1½ hours). Roll dough out to 1/3-inch thickness. Cut with doughnut cutter. Let rise 30-40 minutes. Fry in deep fat.

Icing: Mix together icing ingredients with enough hot water to thin icing. Pour over top of doughnuts.

Mrs. H. F., Hanson Co., S. D.

YEAST BREADS/Rolls & Buns

ICING FOR LONG JOHNS

1 lb. confectioners sugar
1 tbsp. cornstarch
1 tbsp. cream
water

Mix together confectioners sugar, cornstarch, cream and enough water to make a medium consistency. Dip Long Johns while hot. Delicious, too, on raised doughnuts and cinnamon rolls.

Mrs. J. E., Turner Co., S. D.

LONG JOHNS

- 2 pkgs. yeast
- ½ c warm water
- ¾ c milk, scalded and cooled
- ¼ c sugar
- ½ tsp. salt
- 1 c flour
- 1/3 c shortening
- 2 eggs
- 3½ c flour

Soak yeast in warm water for 5 minutes. Add milk, sugar, salt and 1 cup flour. Beat until smooth. Beat in shortening and eggs. Add 3½ cups flour; mix to soft dough. Turn out on lightly floured board. Knead vigorously for 5 minutes. Place in bowl; cover; let rise ½ hour. Roll out. Let rise very lightly. Cut into strips. Fry in deep fat.

Mrs. T. K., Campbell Co., S. D.

POTATO BUNS

- 2 c hot mashed potatoes
- 2 tbsps. sugar
- 1 tbsp. salt.
- 4 tbsps. shortening, melted
- 2 eggs, well beaten
- 1 cake compressed yeast
- 2 c milk
- 6 c flour

Add sugar, salt and shortening to freshly prepared hot mashed potatoes. Cool; add eggs. Soften yeast in milk; add potato mixture. Add flour, a little at a time, beating well after each addition. Turn onto lightly floured board. Knead until smooth and elastic. Dough should be as soft as possible to handle. Cover with damp cloth; set in warm place; allow to double in bulk. Turn onto lightly floured board. Roll in round sheet ½-inch thick. Cut into sections as you would cut a pie, having each section about 3 inches wide at outer edge. Begin at outside, rolling each section toward center. Form into crescents. Place on oiled baking sheet. Cover with warm damp cloth. Let rise until triple in bulk. Bake at 450 degrees for 12 minutes.

Note: Dough may be made into buns. Sprinkle with flour; bake as above.

E. H., Griggs Co., N. D.

FRENCH BUTTER ROLLS

- 1 c milk, scalded and cooled
- ½ c butter, part margarine
- 1 tsp. salt
- ½ c sugar
- 2 yeast cakes
- ¼ c lukewarm water
- 4 eggs, beaten
- 1 tsp. grated lemon rind
- flour

Mix together milk, butter, salt and sugar. Dissolve yeast in lukewarm water. Combine with cooled milk mixture. Add eggs, grated lemon rind and enough flour to make soft dough. Mix well. Place in greased bowl and let rise. Make into roll; let rise again. Bake at 425 degrees for 10 minutes.

Mrs. M. H., Emmons Co., N. D.

CINNAMON ROLLS

- 2 pkgs. yeast
- 2 c lukewarm water
- 1/3 c sugar
- 1 tbsp. salt
- 6½ c flour
- 2 eggs
- ½ c shortening, melted
- butter
- sugar
- cinnamon

Syrup:

- 1 c brown sugar
- ½ c water

Dissolve yeast in lukewarm water. Let stand 2 minutes. Add sugar; mix well. Add salt and 2 cups flour; beat for 2 minutes. Add eggs and melted shortening. Beat 1 minute longer. Add 4½ cups flour. Let rise for 25 minutes. Roll out on floured board. Cover with butter, sugar and cinnamon. Roll up. Cut in ½-inch rolls. Let rise. Prepare syrup by boiling together brown sugar and water. Spoon syrup over tops of rolls. Bake at 375 degrees for 20 minutes, or until golden brown.

Mrs. E. H., Turner Co., S. D.

CARAMEL FOR BOTTOM OF ROLLS

- 3 tbsps. butter, melted
- 1 c brown sugar
- 2 tbsps. water
- pecans

Boil butter, brown sugar and water for 1 minute in heavy skillet. Pour into bottoms of muffin pans. Add pecans. Put in dough.

Variation: Place 1 tablespoon butter, 2 teaspoons brown sugar and 2 drops water in each muffin pan. Arrange pecans; add dough.

Note: After removing rolls from oven, let stand a few minutes before inverting.

Mrs. K. D., Burleigh Co., N. D.

ELEPHANT TRACKS

- 3 c milk, warmed
- 1 lb. butter
- ½ c sugar
- rind of 1 lemon
- ½ tsp. nutmeg
- 4 eggs, beaten
- 3 pkgs. dry yeast
- 2 tsps. salt
- flour
- butter
- sugar
- cinnamon
- nuts (optional)

Heat milk, butter, sugar, lemon rind and nutmeg to warm stage. Add eggs, dry yeast, salt, and enough flour to make soft dough. Let rise once. Roll out dough; spread with butter, sugar and cinnamon. Roll up and cut in ½-inch rolls. Roll each slice out on sugar and cinnamon mixture. Bake at once. (Do not let rise.) Add nuts if desired when rolling out rolls.

Mrs. E. T., Roberts Co., S. D.

SWEET ROLL DOUGH

- 2 pkgs. active dry yeast
- ½ c warm water
- ½ c sugar
- 2 tsps. salt
- 2 eggs
- ½ c shortening
- 1½ c lukewarm milk, scalded and cooled
- 7½ c flour
- 4 tsps. cinnamon
- 1 c sugar

Dissolve yeast and water in large pan. Stir in sugar, salt, eggs, shortening and milk. Add 3½ cups flour. Mix well until smooth. Add remaining flour until dough sticks to sides of bowl (about 4 cups). Turn out on floured board. Knead until smooth and elastic (about 10 minutes). Place in greased pan, turning greased side up. Let rise in warm place free from draft until double in bulk. Punch down; let rise again until almost double. Turn out on lightly floured board. Let rest a few minutes. Divide dough into 2 parts. Roll each portion into oblong shapes, ¼-½-inch thick. Cover with butter; sprinkle with cinnamon and sugar. (For tastier rolls, sprinkle also with ½-¾ cup brown sugar.) Beginning with wide end, roll each section into long roll. Cut each roll into 12 parts. Place in greased jelly roll pan. Cover; let rise until double. Bake at 375 degrees for 25-30 minutes. Turn out of pan at once; let cool.

Note: For caramel rolls, put ¼ cup syrup, 2 tablespoons butter and ½ cup brown sugar in pan; mix. Place rolls on top of mixture. Bake.

Mrs. W. L., Pennington Co., S. D.

wheat

Bread serves as a symbol of all foods. But bread is more than just a symbol. Enriched breads and cereals are so important in the daily diet that they comprise one of the four basic food groups, and a good diet includes four or more servings a day from the breads and cereals group.

Our two states rank high in the production of hard red spring wheat. Flour from hard red spring wheat is superior to all others in bread-making qualities, because it contains high-strength gluten, which permits the dough to expand into large, light, porous loaves. We also lead the nation in the production of durum wheat, which is used in high-quality macaroni, spaghetti and noodles.

YEAST BREADS/Rolls & Buns

BUTTER CRISPIES

- 1 pkg. yeast
- ½ c warm water
- 1½ c sugar
- ½ c shortening
- 1¼ tsps. salt
- 1 c milk, scalded and cooled to lukewarm
- 2 eggs
- 1 tsp. lemon extract
- 6-6½ c sifted flour
- ½ c soft butter
- 4 tsps. cinnamon

Soften yeast in warm water. Combine ½ cup sugar, shortening, salt and milk. Blend in eggs, lemon extract and softened yeast. Gradually add flour to form stiff dough. Knead on well-floured surface until smooth. Place in greased bowl; cover and let rise until light and double in bulk (1-1½ hours). Roll dough on floured surface to 15x12-inch rectangle. Spread half the dough with 2 tablespoons soft butter and ¼ cup sugar. Fold unbuttered side over buttered side; seal edges. Roll out again to 15x12-inch rectangle. Repeat butter and sugar measurements. Spread and roll dough two more times, ending with 15x12-inch rectangle. Combine ¾ cup sugar and cinnamon. Sprinkle half the mixture over dough. Roll up, starting with 15-inch side. Cut into ½-inch slices. Place on well-greased cookie sheet and flatten with palm of hand. Sprinkle with cinnamon and sugar. Let rise in warm place for 15-20 minutes. Bake at 400 degrees for 12-15 minutes, or until golden brown.

Note: While first pans of rolls are baking, place extra rolls on aluminum foil to rise. Transfer foil to cookie sheet; bake.

Mrs. A. L., Carson Co., S. D.

YEAST BREADS/Coffee Cakes

KRINGLES

- 1 pkg. yeast
- 2/3 c warm milk
- 1 tbsp. sugar
- 2 egg yolks
- ½ c margarine
- ½ c lard
- ¼ c sugar
- 2½ c sifted flour
- salt
- 2 egg whites
- ¼ c sugar

Dissolve yeast in mixture of warm milk, sugar and egg yolks. Combine margarine, lard ¼ cup sugar, flour and salt. Add yeast mixture. Divide dough into 4 parts. Roll out 2 parts to fit 10x15-inch pan. Place filling on top of each. Roll out other 2 parts; place on top filling. Beat egg whites and ¼ cup sugar until stiff. Spread over kringles. Let rise for 1 hour Bake at 350 degrees for 20-30 minutes.

Mrs. A. H., Pipestone Co., Minn.

NUT-FRUIT COFFEE CAKE

2 pkgs. active dry or cake yeast
½ c warm water
½ c sugar
½ c butter or margarine, melted
2 tsps. salt
1 c milk, scalded
1 egg
1 egg yolk
6½-7 c sifted flour
1½ c apricot preserves
1 tbsp. lemon juice
½ tsp. ground cardamom
1 c chopped walnuts
6 tbsps. orange marmalade
1 egg, beaten
Pecan-Raisin Filling:
1¼ tsps. cinnamon
¼ c sugar
1½ tsps. grated lemon rind
½ c seedless raisins
½ c light cream

Sprinkle or crumble yeast into warm water to soften. Combine sugar, butter, salt and milk. Stir in 1 egg yolk and softened yeast mixture. Gradually stir in enough flour to form stiff dough. On floured board, knead dough until smooth (5 minutes). Place in greased bowl, cover with towel and let rise in warm place (80-85 degrees) until double in bulk (about 1¼ hours). Blend together apricot preserves, lemon juice and cardamom. When dough has doubled, punch it down; divide a little more than half of it into 2 parts. On floured board, roll 1 part into 12-inch square. Over it spread half the apricot preserve mixture; sprinkle with ½ cup walnuts. Roll up, jelly-roll fashion. Reapeat with other part, using remaining preserves and walnuts. Place rolls, side by side, in greased 12x8x2-inch baking dish; flatten lightly With scissors, snip out 6 small V's down top of each, lifting each point up and back. Let rise, uncovered, in warm place until doubled in bulk. Roll remaining dough into rectangle 17x8 inches. Over it, evenly spread orange marmalade, then pecan-raisin filling. Starting with 8-inch ends, roll up towards middle from both ends. Place in greased 10x5x3-inch loaf pan. With scissors, snip out 4 small V's on each strip, down top of cake, lifting each point up and back. Cover; let rise in warm place until doubled. Brush tops of both cakes with 1 beaten egg. Bake at 350 degrees for 35-40 minutes, or until golden brown.

Pecan-Raisin Filling: Combine filling ingredients.

Mrs. A. G., Kidder Co., N. D.

sugar

Sugar is a necessary ingredient in every kitchen. The sugar on our tables and in our cakes comes from two sources — sugar cane and sugar beets. The growing of sugar beets is important to the economy of the Dakotas, especially in the Red River Valley area of North Dakota. Insist that the sugar you buy be produced from Dakota-grown beets.

BROWN BREAD

- 4 tbsps. butter
- ½ c sugar
- 1 egg
- ½ c light molasses
- 2 c flour
- 2 tsps. baking powder
- ½ tsp. soda
- 1 tsp. salt
- 1 c whole wheat flour
- 1½ c buttermilk
- ½ c raisins
- ½ c chopped nuts

Mix ingredients in order given. Bake in bread loaf pan at 350 degrees for about 1 hour, or until done.

Mrs. H. B., Grand Forks Co., N. D.

APPLESAUCE BREAD

- ½ c shortening
- 2/3 c brown sugar
- 2 eggs
- 2 c sifted flour
- 1 tsp. baking powder
- 1 tsp. soda
- ½ tsp. salt
- 1 c thick applesauce
- ½ c nuts

Icing:

- ½ c confectioners sugar
- 1 tbsp. warm water
- dash of salt

Cream together shortening and brown sugar. Add eggs; beat well. Sift together flour, baking powder, soda and salt; add alternately with applesauce to creamed mixture. Stir in nuts. Pour into greased 5x10x3-inch loaf pan. Bake at 350 degrees for 50-55 minutes.

Icing: Mix together icing ingredients. Frost loaf while warm, allowing frosting to run down sides for decorative effect.

Mrs. K. K., Clay Co., S. D.

BANANA NUT BREAD

- ½ c butter
- 1 c sugar
- ½ tsp. salt
- 2 egg yolks, beaten
- 2 c flour
- 4 tsps. baking powder
- 1 c milk
- 1 c mashed bananas
- ½ c chopped nuts
- 2 egg whites, stiffly beaten

Cream together butter, sugar and salt. Add egg yolks; mix slightly. Combine flour and baking powder; add alternately with milk to creamed mixture. Add banana pulp and nuts; mix well. Fold in egg whites. Place in buttered bread pan. Bake at 350 degrees for 1 hour.

Variation: Use ½ cup whole wheat flour instead of 2 cups flour, and use sour milk instead of sweet with 3 teaspoons baking powder and ½ teaspoon soda.

Mrs. W. B. H., Stevens Co., Minn.

PEANUT BREAD

- 1 egg
- 1 c sugar
- 2 c flour
- 2 tsps. baking powder
- ½ tsp. salt
- 1 c milk
- ½ c chopped peanuts

Beat together egg and sugar. Add sifted dry ingredients alternately with milk. Stir in peanuts. Bake in loaf pan at 350 degrees for 40 minutes or until toothpick comes out clean.

Mrs. J. G. B., Turner Co., S. D.

APRICOT NUT LOAF

- 3 c sifted flour
- 3 tsps. baking powder
- ¾ tsp. salt
- ½ c brown sugar
- ½ c finely broken walnuts
- 1 c finely cut, dried apricots
- 1 egg, well-beaten
- 1 c milk
- 4 tbsps. butter, melted

Sift together sifted flour, baking powder and salt. Add sugar; mix well. Add nuts and apricots. Combine egg, milk and shortening; add to flour mixture; blend. Bake in greased 8x4x3-inch loaf pan at 350 degrees for 1 hour or until done.

Ms. T. E. M., Hand Co., S. D.

CRANBERRY HOLIDAY BREAD

- 2 c coarsely chopped cranberries
- 1 c chopped walnuts
- 1 c chopped dates
- ¾ c diced, mixed candied fruit
- ¼ c flaked coconut
- 4 c sifted flour
- 2 c sugar
- 1 tsp. salt
- 3 tsps. baking powder
- 1 tsp. soda
- 2 eggs, beaten
- grated rind of 2 oranges
- 1 c orange juice
- ¼ c hot water
- ¼ c butter, melted

Mix together chopped cranberries, walnuts, dates, candied fruit and coconut; set aside. Sift together dry ingredients. Combine eggs, orange juice and hot water. Add to flour mixture, quickly mixing until dry ingredients are moistened. Stir in fruit mixture, a fourth at a time. Fold in melted butter. Spread batter into two 9x5x3-inch loaf pans, greased on bottoms only. Bake at 325 degrees for 1 hour 10 minutes, or until done. Remove from pans at once; cool.

Note: This freezes well and improves with age.

Mrs. V. B., Sheridan Co., N. D.

QUICK BREADS

DATE AND NUT BREAD

- 1 (12-oz.) pkg. dates, cut fine
- 2 tsps. soda
- 2 c boiling water
- 2 c sugar
- 2 tbsps. shortening
- 2 eggs
- 4 c flour
- 1 tsp. salt
- 1 c chopped nuts
- 1 tsp. soda

Sprinkle 2 tsps. soda over cut dates. Add boiling water; set aside to cool to lukewarm. Cream together sugar and shortening; add eggs. Beat well. Sift flour with salt; add to creamed mixture. Fold in date mixture. Add nuts and 1 teaspoon soda; mix together. Grease and flour 12 soup cans. Fill half full. Bake at 350 degrees for 1 hour.

Mrs. M. H., Campbell Co., S. D.

HOBO BREAD

- 1 c boiling water
- 2 tsps. soda
- 1½ c raisins
- 1 c sugar
- 2 tbsps. shortening
- 1 egg
- ½ tsp. salt
- ½ tsp. vanilla
- 2 c flour

Mix together water, soda and raisins; cool. Combine sugar, shortening, egg, salt, vanilla and flour; mix well. Add raisin mixture. Put into loaf pans or round tin cans. Bake at 350 degrees for 35-40 minutes.

Variation: Try adding 2 tablespoons molasses and chopped nuts for a different flavor.

Mrs. A. H., Grand Forks Co., N. D.

PUMPKIN BREAD

- 1½ c sugar
- ½ c cooking oil
- 2 eggs
- 1 c pumpkin
- 1¾ c flour
- ¼ tsp. baking powder
- 1 tsp. soda
- 1 tsp. salt
- ½ tsp. cloves
- ½ tsp. cinnamon
- ½ tsp. nutmeg
- ½ tsp. allspice
- 1/3 c water
- ½ c raisins
- ½ c crushed walnuts

Cream together sugar and oil. Add eggs and pumpkin. Sift together dry ingredients; add to pumpkin mixture; mix well. Add water, raisins and walnuts. Bake at 350 degrees for 1 hour.

Mrs. J. R., Emmons Co., N. D.

PINEAPPLE NUT BREAD

2 c sifted flour
½ c sugar
1 tsp. baking powder
½ tsp. salt
1 c raisins
½ c coarsely chopped walnuts
1 egg, beaten
1 tsp. vanilla
2 tbsps. shortening, melted
1 tsp. soda
1 (No. 1) can crushed pineapple, undrained

Sift together flour, sugar, baking powder and salt. Add raisins and nuts. Combine egg, vanilla and shortening; add to mixture. Dissolve soda in pineapple; add to mixture. Stir until blended. Pour into greased 8x4x4-inch loaf pan. Bake at 350 degrees for 1 hour, or until done. Cool on rack. Serve buttered or spread with cream cheese.

Ms. L. B., Pennington Co., S. D.

SODA BREAD

2 c flour
4 tsps. soda
1 tsp. sugar
1 tsp. salt
½ c raisins
½ c currants
1 tbsp. caraway seed
3 tbsps. shortening
2/3 c milk

Mix together flour, soda, sugar, salt, raisins, currants and caraway seed. With a pastry blender or fork, work in shortening. Quickly stir in milk. Add more milk, little by little, until dough is soft and light but not sticky. Turn out onto floured board. Knead with floured hands for about 20 strokes or until smooth. Spread in greased heavy frying pan or 9-inch round tin. Bake at 350 degrees for 30 minutes; increase heat to 400 degrees and bake 5 minutes longer.

Mrs. P. F., Brown Co., S. D.

MOLASSES CORN BREAD

1 c flour
1 c corn meal
1 tsp. salt
4 tsps. baking powder
1 egg, well-beaten
¼ c molasses
1 c milk
¼ c shortening, melted

Mix together dry ingredients. Combine egg, molasses and milk; stir into dry mixture. Stir in shortening. Pour into a 9-inch square pan; bake at 400 degrees for 25 minutes

Ms. E. H., Brown Co., S. D.

QUICK BREADS

LEMON TEA BREAD

1 c sugar
½ c shortening
2 eggs, slightly beaten
1¾ c flour
½ tsp. salt
1 tsp. baking powder
½ c milk
grated rind of 1 lemon
½ c chopped walnuts
½ c sugar
juice of 1 lemon

Cream together sugar and shortening; add eggs. Sift together flour, salt and baking powder; add alternately with milk to shortening mixture, stirring constantly. Mix in lemon rind and chopped nuts. Bake in 9x5-inch loaf pan at 350 degrees for 1 hour. When golden brown pour lemon topping (sugar and lemon juice) over hot bread while still in pan. Leave in pan for 10 minutes before removing. Cool, slice and serve.

Mrs. V. S., Wilkin Co., Minn.

CARROT BREAD

2 eggs, beaten
1 c sugar
¾ c oil
1½ c flour
½ tsp. salt
1 tsp. soda
1 tsp. cinnamon
2 c finely grated raw carrots
½ c fine coconut

Mix together eggs and sugar. Add oil; beat well. Add remaining ingredients; mix well. Bake at 325 degrees for 45 minutes.

Ms. M. F., Cass Co., N. D.

CORN BREAD

1 c flour
3½ tsps. baking powder
1 tsp. salt
3 tbsps. sugar
1 c yellow corn meal
1 egg, slightly beaten
1 c milk
¼ c cooking oil

Sift together flour, baking powder, salt, sugar and corn meal. Add remaining ingredients. Stir until flour is moistened (do not beat). Bake at 425 degrees for 20-25 minutes.

Mrs. A. A. B., Hutchinson Co., S. D.

DUMPLINGS

- 1 c flour
- 3 tsps. baking powder
- 2 eggs, slightly beaten
- salt
- sugar
- 2 tbsps. cream

Sift together flour and baking powder. Mix together eggs, salt and sugar to taste. Add cream to egg mixture; gradually add flour mixture. Drop into boiling water.

Mrs. D. K., Kingsbury Co., S. D.

POTATO DUMPLINGS

- 6 c grated potatoes
- 6 tbsps. sweet cream
- 1 tsp. baking powder
- 1 tsp. salt
- ½ c flour

Drain grated potatoes; mix with remaining ingredients. Roll by hand into walnut-size balls. Drop each dumpling in boiling salt water. Boil until done. Drain in colander. Serve with white sauce.

Mrs. W. H., Spink Co., S. D.

HOME BISCUIT MIX

- 9 c flour
- ¼ c sugar
- 1/3 c baking powder
- 1 tbsp. salt
- 2 tsps. cream of tartar
- 2 c lard

Mix together all ingredients in 5-quart ice cream bucket; cover; store in cupboard without refrigeration.

Biscuits: Add 1 cup milk to 3 cups mix.

Muffins: Add 2 tablespoons sugar, 1 egg and 1 cup milk to 3 cups mix.

Waffles: Add 1 egg and 1½ cups milk to 3 cups mix.

Mrs. N. J., Harding Co., S. D.

BAKING POWDER BISCUITS

- 2 c flour
- 3 tsps. baking powder
- 1 tsp. salt
- ¼ c shortening
- milk

Sift together dry ingredients. Cut in shortening until mixture looks like meal. Stir in ½ cup milk; mix well. Add additional milk until dough is soft and easy to roll out. Place dough on floured board; knead lightly about 20 times. Roll out to ½-inch thick. Cut with floured biscuit cutter or medium size glass, floured. Place biscuits close together for biscuits with soft sides, or about an inch apart for biscuits with crusty sides. Place on ungreased baking sheet; bake at 450 degrees for 10-12 minutes.

Mrs. W. L., Pennington Co., S. D.

BLUSHING PEAR COFFEE CAKE

Fruit Mixture:
¼ c sugar
¼ tsp. allspice
¼ tsp. cloves
2 tbsps. cinnamon candies
2 tbsps. water
2 tbsps. butter
2 pears, peeled and sliced

Batter:
1½ c sifted flour
2 tsps. baking powder
½ tsp. salt
½ c sugar
1 egg, beaten
½ c milk
3 tbsps. shortening, melted
4 pears, diced

Fruit Mixture: Combine sugar, allspice, cloves and cinnamon candies in small saucepan. Add water; heat to boiling. Continue boiling about 3 minutes, stirring often. Remove from heat; add butter and cool. Arrange pears in bottom of greased 8x8-inch pan. Pour syrup mixture evenly over pears.

Batter: Sift together flour, baking powder, salt and sugar. Combine egg, milk and shortening. Add to flour mixture; stir until smooth. Pour over fruit mixture; bake at 400 degrees for 25 minutes.

Ms. J. M., Codington Co., S. D.

SOUR CREAM COFFEE CAKE

1 c margarine
2 eggs
2 c sugar
1 c sour cream
½ tsp. vanilla
1¼ tsps. baking powder
2 c cake flour
½ tsp. salt

Topping:
½ c finely chopped nuts
2 tsps. brown sugar
½ tsp. cinnamon
confectioners sugar

Cream together margarine, eggs and sugar. Fold in sour cream. Add vanilla, baking powder, cake flour and salt. Prepare topping by combining nuts, brown sugar and cinnamon. Spread half of batter in greased tube pan. Sprinkle half topping mixture on batter. Add remaining batter; sprinkle remainder of topping on top. Bake at 350 degrees for 1 hour. Sprinkle with confectioners sugar while warm.

Mrs. W. R., Mercer Co., N. D.

BLUEBERRY BUCKLE

¾ c sugar
¼ c soft shortening
1 egg
¼ c milk
2 c flour
2 tsps. baking powder
½ tsp. salt
2 c blueberries, fresh or drained canned
½ c sugar
1/3 c flour
1 tsp. cinnamon
¼ c soft butter

Mix thoroughly ¾ cup sugar, shortening and egg. Stir in milk, Combine 2 cups flour, baking powder and salt; add to sugar mixture. Blend in berries. Spread in greased and floured 9-inch square pan. Combine ½ cup sugar, 1/3 cup flour, cinnamon and butter; sprinkle over cake mixture. Bake at 375 degrees for 45-50 minutes, or until toothpick comes out clean. Serve warm.

Ms. K. S., Wilkin Co., Minn.

HONEY CRISP COFFEE CAKE

1½ c sifted flour
2 tsps. baking powder
½ tsp. salt
½ c sugar
1 egg, beaten
½ c milk
3 tbsps. shortening, melted
3 tbsps. softened butter or margarine
1/3 c honey
¼ c shredded or flaked coconut
½ c crushed corn flakes
½ c crushed pineapple, drained

Sift together flour, baking powder, salt and sugar. Combine egg, milk and shortening. Add to flour mixture; stir until smooth. Pour into greased 8-inch square pan. Cream together butter or margarine and honey until light and fluffy. Add coconut, corn flakes and pineapple; mix thoroughly. Spread this mixture over coffee cake batter. Bake at 400 degrees for 25 minutes.

Mrs. T. K., Adams Co., N. D.

AFTERNOON COFFEE CAKE

1 c sugar
1 egg, beaten
1 c sour cream
¼ tsp. salt
¼ tsp. nutmeg
2 c sifted flour
1 tsp. soda
1¾ tsps. baking powder
1 c diced raw apple
¼ c sugar
1 tsp. cinnamon
1 tbsp. butter, melted
chopped nuts (optional)

Mix together 1 cup sugar, egg, sour cream, salt, nutmeg, flour, soda and baking powder. Fold in diced apple, a little at a time. Pour into pan; top with ¼ cup sugar, cinnamon, melted butter and chopped nuts. Bake at 350 degrees for 30 minutes.

Mrs. E. H., Turner Co., S. D.

QUICK BREADS/Coffee Cakes

STREUSEL-FILLED COFFEE CAKE

¾ c sugar
¼ c shortening
1 egg
½ c milk
1½ c sifted flour
2 tsps. baking powder
½ tsp. salt

Streusel Mixture:
½ c brown sugar
2 tbsps. flour
½ c chopped nuts
2 tsps. cinnamon
2 tbsps. butter, melted.

Mix together sugar, shortening and egg, Stir in milk. Sift together flour, baking powder and salt; add to sugar mixture. Spread half of the batter in well-greased, floured 9-inch square pan. Prepare streusel mixture by combining brown sugar, flour, nuts, cinnamon and melted butter; mix well. Sprinkle batter with half of streusel mixture. Add remaining batter; sprinkle remaining streusel over top. Bake at 350 degrees for about 30 minutes.

Mrs. E. H., Turner Co., S. D.

JAM STREUSEL COFFEE CAKE

1½ c sifted flour
2 tsps. baking powder
½ tsp. salt
½ c sugar
1 egg, beaten
½ c milk
3 tbsps. shortening, melted
2 tbsps. butter or margarine
¼ c sugar
¼ c flour
¼ c jam

Sift togther 1½ cups flour, baking powder, salt and ½ cup sugar. Combine egg, milk and shortening. Add to flour mixture; stir until smooth. Pour into greased 8-inch square pan. Combine butter or margarine, ¼ cup sugar and ¼ cup flour. Mix until crumbly. Sprinkle this mixture over batter to form squares. Fill squares with jam. Bake at 400 degrees for 25 minutes.

Mrs. E. J. K., Brookings Co., S. D.

QUICK BREADS/Doughnuts

DROP DOUGHNUTS

3 eggs, beaten
1 c sugar
1 c milk
½ c sweet cream
3½ c flour
2 tsps. baking powder
mace or vanilla

Combine eggs, sugar, milk and cream. Flavor to suit your taste with mace or vanilla. Sift together flour and baking powder; add to egg mixture. Drop from teaspoon into deep hot fat; fry until nice golden brown.

Mrs. E. H., Sheridan Co., N. D.

FAMILY DOUGHNUTS

- 4 eggs, lightly beaten
- 2 c sugar
- 1 / 3 c lard, melted
- 2 tsps. baking soda
- 2 c buttermilk
- 1 tsp. nutmeg
- 1 tsp. salt
- 2 tsps. baking powder
- 5 c flour

Beat eggs and sugar until creamy; add melted lard. Combine baking soda and buttermilk. Sift together remaining ingredients. Alternately add flour mixture with buttermilk mixture to sugar mixture. Blend well.

Note: This dough can be used in doughnut maker, or add more flour and roll dough.

Mrs. B. S., Sargent Co., N. D.

APPLE DOUGHNUTS

- 1 c sifted flour
- ½ tsp. salt
- 1½ tsp. baking powder
- 1 tbsp. sugar
- 1 egg, beaten
- ½ c milk
- 1 tbsp. fat or oil, melted
- 1 c raw, thinly sliced apples
- dash nutmeg or cinnamon

Sift together dry ingredients. Combine egg, milk and fat; add to dry ingredients all at once. Blend only to moisten. Stir in apples. Drop by spoonfuls into hot fat (350-375 degrees). Cook until brown. Drain on toweling. Place in paper bag with sugar and cinnamon; shake; cool.

Mrs. E. M., Davison Co., S. D.

POTATO DOUGHNUTS

- 2 eggs, beaten
- 1½ c sugar
- 1 c hot mashed potatoes
- 2 tbsps. butter, melted
- ½ c milk
- ½ tsp. salt
- 3 c flour
- 3 tsps. baking powder
- ⅛ tsp. salt
- ⅛ tsp. nutmeg
- ½ tsp. cinnamon

Mix together eggs and sugar. Mix together potatoes, butter, milk and salt; add to egg mixture. Sift together remaining ingredients; add to potato mixture. Let stand. Roll to 3/8-inch thickness. Cut with doughnut cutter; fry in hot fat.

Miss G. B., Moody Co., S. D.

SWEET POTATO DOUGHNUTS

5 c flour
7 tsps. baking powder
1½ c warm mashed sweet potatoes (or pumpkin, squash or carrots)
1 tsp. salt
3 eggs
2 c sugar
2 tbsps. butter, melted
1 tsp. nutmeg
1 c milk
confectioners sugar

Sift together flour and baking powder. Whip potatoes (or potato substitute); gradually add salt, eggs, sugar, butter, nutmeg and milk; blend thoroughly. Add flour mixture; mix well. Chill for at least 1 hour. Turn dough onto floured board. Roll or pat to ½-inch thickness. Cut with doughnut cutter; drop into 375 degree fat. Turn doughnuts as they brown. Drain; sprinkle with confectioners sugar; serve hot.

Mrs. E. S., Stutsman Co., N. D.

SOUR MILK DOUGHNUTS

2 eggs
1 c sugar
¾ c thick sour milk
2 tbsps. thick sweet cream
2 tbsps. butter, melted
2 c sifted flour
2 tsps. baking powder
1 tsp. soda
½ tsp. nutmeg
flour

Mix together eggs, sugar, sour milk, cream and melted butter. Sift together 2 cups sifted flour, baking powder, soda and nutmeg. Add to milk mixture; mix well. Add enough flour to make soft dough. Chill 2 hours or more. Roll into ¼-inch thickness; cut with doughnut cutter; fry in deep fat.

Mrs. A. S., Barnes Co., N. D.

BANANA DOUGHNUTS

5 c flour
4 tsps. baking powder
1 tsp. soda
2 tsps. salt
1 tsp. nutmeg
¼ c shortening
1 c sugar
3 eggs, beaten
¾ c mashed bananas
½ c sour milk
1½ tsps. vanilla

Sift together flour, baking powder, soda, salt and nutmeg. Cream shortening; slowly add sugar. Add eggs; beat vigorously. Stir in mashed bananas, sour milk and vanilla. Add dry ingredients. Knead lightly. Roll out to ½-inch thickness. Fry in deep fat for about 3 minutes.

Mrs. D. S., Corson Co., S. D.

SPEEDY CEREAL DOUGHNUTS

1 (8-oz.) pkg. refrigerated buttermilk biscuits
1 c sifted confectioners sugar
2 tbsps. orange juice
1 c presweetened puffed oat cereal, crushed

Push a small hole through the center of each biscuit. Fry in 1 inch of hot fat (360 degrees) until golden brown on both sides, turning once. Drain on toweling. Combine sugar and orange juice; mix well. Dip hot doughnuts into glaze; roll in cereal crumbs. Serve while warm.

Mrs. R. K., Foster Co., N. D.

Muffins/QUICK BREADS

HONEY MUFFINS

1 c milk
½ c honey
2 tbsps. sugar
3 eggs, beaten
¼ c butter, melted
3 c flour
6 tsps. baking powder
¾ tsp. salt

Combine milk, honey and sugar; add eggs and butter. Combine remaining ingredients; add to liquid mixture. Stir quickly. Mixture should be lumpy. Put into greased muffin pans; bake at 375 degrees for 12-15 minutes.

Mrs. N. E., Cass Co., N. D.

SOUR CREAM RAISIN MUFFINS

1 c sifted flour
½ c sugar
2½ tsps. baking powder
½ tsp. salt
¼ tsp. soda
1 c rolled oats
3 tbsps. shortening, melted
1 egg, beaten
½ c sour cream
½ c milk
½ c raisins

Sift together flour, sugar, baking powder, salt and soda. Stir in rolled oats. Add remaining ingredients. Mix only until dry ingredients are moistened. Fill greased muffin cups 2/3 full. Bake at 425 degrees for 12-15 minutes, or until done.

Mrs. E. S., Logan Co., N. D.

OATMEAL MUFFINS

1 c oatmeal
1 c buttermilk or sour milk
1 egg
½ c shortening, melted
½ c brown sugar
1 c flour
½ tsp. salt
1 tsp. baking powder
½ tsp. soda

Soak oatmeal in buttermilk or sour milk for 1 hour. Add egg; beat well. Add melted shortening and brown sugar. Sift together remaining ingredients; add to oatmeal mixture. Bake in greased muffin pans at 400 degrees for 20 minutes.

Mrs. H. S., Brown Co., S. D.

MUFFINS

½ c sugar
½ c shortening
1 egg
1½ c flour
2 tsps. baking powder
½ tsp. salt
½ c milk

Cream together sugar and shortening; add egg and beat well. Sift together flour, baking powder and salt; add alternately with milk to creamed mixture. Blend well, but do not beat. Fill well-greased muffin tins 2/3 full. Bake at 400 degrees for 20-25 minutes.

Variation: Add ½-1 cup fresh or frozen blueberries; or add 1 teaspoon grated orange rind and sprinkle tops with sugar; or add ½ cup currants and sprinkle tops with cinnamon and sugar.

Mrs. H. S., Brown Co., S. D.

ORANGE-RAISIN MUFFINS

2 c sifted flour
¾ tsp. soda
½ tsp. salt
¼ c sugar
½ c raisins
1 tsp. grated orange rind
1 egg, beaten
¾ c buttermilk
¼ c orange juice
¼ c shortening, melted

Sift together flour, soda, salt and sugar. Stir in raisins and orange rind. Combine remaining ingredients; add to flour mixture, stirring only until flour is moistened. Fill greased muffin pans 2/3 full. Bake at 425 degrees for 20 minutes.

Miss J. L. M., Hand Co., S. D.

JEWISH POTATO PANCAKES

- 2 lbs. potatoes
- 2 tbsps. self-rising flour
- 1 egg, beaten
- salt

Peel and finely grate potatoes into a bowl of cold water. Strain through a muslin bag. Place strained potatoes in bowl; add flour, egg and salt. Mix thoroughly. Drop by tablespoons into hot oil; fry on both sides until nicely browned.

Note: The secret to making these pancakes is to drain all excess liquid from grated potatoes.

Mrs. W. H., Brookings Co., S. D.

BUTTERMILK WAFFLES

- 2 c sifted flour
- 2 tsps. baking powder
- ½ tsp. salt
- ½ tsp. soda
- 3 tbsps. sugar
- 2 eggs, beaten
- 2 c buttermilk
- ¼ c shortening, melted

Sift together dry ingredients. Combine eggs, buttermilk and shortening. Beat all together; bake in waffle iron.

Mrs. R. R. H., LaMoure Co., N. D.

CAKES & FROSTING

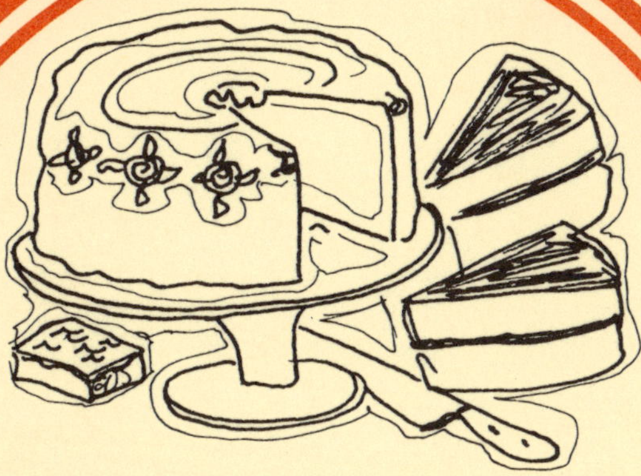

Cookies may be the children's favorite and pies please the men, but it's with cakes that a woman displays her most creative talents. It's not unusual for a homemaker to earn her reputation as a "good cook" on the basis of a cake and be asked over and over to bring one of her special cakes to the church bake sale or the club supper. Try the cake recipes on these pages and stake your "claim to fame" as a master cook.

SOUR CREAM DEVILS FOOD CAKE

½ c cocoa
1½ c boiling water
1 c brown sugar
1 c sugar
1 c sour cream
2 eggs
½ tsp. salt
2 tsps. soda
2½-3 c flour

Mix together cocoa and boiling water. Set aside. Mix sugars, sour cream and eggs. Beat well. Combine dry ingredients; add to sour cream mixture; beat well. Add cocoa mixture. Mix and pour into well-greased 9x13 cake pan. Bake at 325 degrees for 45 minutes.

Mrs. A. P., McCook Co., S. D.

CHOCOLATE CHIP CAKE

2 2/3 c flour
1 7/8 c sugar
4½ tsps. baking powder
1 tsp. salt
2/3 c shortening
1¼ c milk
2 tsps. vanilla
5 egg whites
3 sqs. finely shaved sweet or semi-sweet chocolate

Sift together flour, sugar, baking powder and salt. Cream shortening, milk and vanilla. Add dry ingredients; mix at medium speed 2 minutes. Add egg whites; beat 2 minutes. Carefully fold in chocolate. Bake in 9x13-inch pan at 350 degrees for 30-35 minutes.

Note: A potato peeler works well to shave chocolate.

Mrs. B. P., Lake Co., S. D.

CHOCOLATE SPICE CAKE

2 c flour
2 c sugar
1 c margarine
4 tbsps. cocoa
1 c water
½ c sour milk
2 eggs
½ tsp. cinnamon
1 tsp. soda
¼ tsp. salt
1 tsp. vanilla

Frosting:
¼ c margarine
¼ c milk
1 c sugar
½ c chocolate chips

Blend flour and sugar; set aside. Bring to boil the margarine, cocoa and water. Pour cocoa mixture over flour and sugar; blend well. Add remaining cake ingredients; mix well. Pour into greased 15½x10½-inch cookie sheet. Bake at 325 degrees for 30 minutes.

Frosting: Bring to boil margarine, milk and sugar. Add chocolate chips; beat with mixer. Do not cool. Pour over warm cake.

Mrs. L. R., Edmunds Co., S. D.

CAKES/Chocolate

EASY GERMAN CHOCOLATE CAKE

1 pkg. white cake mix
1 pkg. instant chocolate pudding mix
2 c milk
2 eggs
Frosting:
1 c cream
¼ c margarine
2 c sugar
2 egg yolks
1 tsp. vanilla
1 c coconut
1 c chopped pecans

Combine first four ingredients; mix well. Pour into greased and floured 9x13-inch pan. Bake at 350 degrees for 45 minutes.

Frosting: Cook until thick (about 12 minutes) cream, margarine, sugar and egg yolks. Add remaining ingredients. Beat until thick enough to spread.

Mrs. J. G., Sully Co., S. D.

CAKES/Fruit

GRANNY'S FRUIT CAKE

1 c butter
1 c sugar
4 egg yolks, well-beaten
2 c flour
2 tsps. baking powder
1/3 tsp. nutmeg
1 tsp. ground cloves
2 tsps. allspice
1/3 tsp. ground cinnamon
½ c boiled-down cider
1 c citron, diced
1 c candied cherries, diced
2 c seedless raisins
1 c currants
1/6 lb. candied pineapple, diced
1/6 lb. candied orange peel, diced
1/6 lb. candied lemon peel, diced
½ lb. figs, chopped
1 c chopped, blanched almonds
1 c seeded dates, chopped
4 egg whites, stiffly beaten

Cream butter and sugar together. Add egg yolks; blend well. Sift flour, baking powder and spices together. Take out 1 cup of the flour mixture. Boil down cider by boiling 1 cup cider until only ½ cup remains. Add flour mixture and boiled-down cider alternately to creamed mixture. Mix fruits and nuts with reserved flour, coating well; add to cake batter. Fold in egg whites. Pour into a 9-inch tube pan, which has been greased, lined with heavy waxed paper and greased again. Bake at 270 degrees for 4 hours or until firm.

Mrs. J. N., Kingsbury Co., S. D.

AMALGEMASIN CAKE

2 c sugar
1½ c butter
6 eggs
1 tsp. soda
1 c sour milk
4 c flour
1 tsp. baking powder
1 tsp. cloves
1 tsp. cinnamon
1 tsp. allspice
2 c blackberry jam
1 c ground raisins

Cream together sugar and butter. Stir in eggs, one at a time. Dissolve soda in buttermilk; add to creamed mixture. Sift together dry ingredients; add to batter. Fold in jam and raisins. Line 4 8x8-inch pans with waxed paper; do not grease sides. Bake at 325 degrees for 45 minutes or until done.

Mrs. L. P., Roberts Co., S. D.

CARROT CAKE

2 c flour
2 tsps. soda
2 tsps. cinnamon
2 c sugar
1½ c vegetable oil
4 eggs
3 c grated carrots
Frosting:
4 ozs. cream cheese
3 c confectioners sugar
¼ c margarine
2 tsps. vanilla

Mix together dry ingredients; add vegetable oil; mix well. Add eggs, one at a time; beat. Add carrots. Bake at 350 degrees for 1 hour.

Frosting: Beat together frosting ingredients; frost.

Mrs. J. N., Dickey Co., N. D.

MINCEMEAT CAKE

2 c sifted flour
¾ c sugar
2½ tsps. baking powder
¾ tsp. salt
1/3 c shortening
1 egg, well beaten
½ c milk
¾ c moist mincemeat
Glaze:
2 tbsps. butter
3 tbsps. hot milk
1 c confectioners sugar
⅛ tsp. salt

Sift together flour, sugar, baking powder and salt. With pastry blender, cut in shortening until mixture resembles coarse corn meal. Blend together egg, milk and mincemeat. Add to flour mixture; stir until dry ingredients are moistened. Pour into well-greased 1½-quart ring mold. Bake at 375 degrees for 30-35 minutes. Cool in mold on wire rack 10 minutes. Remove from mold and while still warm, drizzle glaze over top and sides of cake.

Glaze: Combine butter in hot milk; stir until butter is melted. Add confectioners sugar and salt; stir until smooth. Stir in more confectioners sugar if a thicker glaze is desired.

Mrs. L. B., Haakon Co., S. D.

MAPLE NUT CAKE

½ c butter
1 c brown sugar
2 egg yolks
2 c flour
1 tsp. salt
2 tsps. baking powder
1 c milk
1 c chopped nuts
2 egg whites, beaten
1 tsp. maple extract
1 tsp. vanilla

Cream together butter and sugar. Add egg yolks; mix well. Sift together flour, salt and baking powder; add to creamed mixture alternating with milk. Add nuts; fold in beaten egg whites. Add maple extract and vanilla. Bake at 350 degrees for about 40 minutes, or until done. Frost with brown sugar frosting flavored with maple extract.

Mrs. D. P., Renville Co., N. D.

PEANUT BUTTER CUPCAKES

½ c peanut butter
1/3 c shortening
1 tsp. vanilla
1½ c brown sugar
2 eggs
2 c sifted flour
2 tsps. baking powder
½ tsp. salt
¾ c milk

Frosting:
1 c confectioners sugar
¼ tsp. banana flavoring
1 tbsp. butter, melted
cream

Cream together peanut butter, shortening and vanilla. Gradually add brown sugar; beat until light and fluffy. Add eggs one at a time, beating well after each. Sift together dry ingredients; add alternately with milk. Place paper baking cups in muffin pans; fill ½ full. Bake at 375 degrees for 20 minutes or until done.

Frosting: Combine confectioners sugar, banana flavoring, butter and just enough cream to make good spreading consistency.

Mrs. H. R., McLean Co., N. D.

FRUIT COCKTAIL CAKE

1½ c sugar
2 c sifted flour
2 tsps. soda
2 eggs, beaten
1 (No. 2) can fruit cocktail
½ c raisins
1 c brown sugar
½ c chopped nuts

Sift together sugar, flour and soda. Add eggs, fruit cocktail, undrained, and raisins; mix well. Pour into a greased and floured 9x13-inch pan. Top with brown sugar and nuts. Bake at 300 degrees for 1 hour and 10 minutes. Serve with vanilla ice cream or whipped cream sweetened with confectioners sugar and almond extract.

Mrs. R. B., Hutchinson Co., S. D.

DATE CAKE

1 c boiling water
1½ c finely chopped dates
½ tsp. soda
½ c shortening
1 c sugar
2 eggs, well beaten
1½ c flour
1 tsp. baking powder
1 tsp. vanilla

Pour boiling water over dates and soda; let cool. Cream together shortening and sugar; add eggs and vanilla. Add dry ingredients to creamed mixture; mix well. Add date mixture. Bake at 350 degrees for 30-40 minutes.

Mrs. P. G., Walworth Co., S. D.

Spice/CAKES

PUMPKIN GINGER BREAD

½ c shortening
½ c sugar
1 egg
1 c sour milk
½ c pumpkin
½ c dark molasses
2¼ c sifted flour
1 tsp. soda
½ tsp. salt
1 tsp. ginger
1 tsp. cinnamon

Cream shortening, sugar and egg. Blend sour milk, pumpkin and molasses; add alternately to creamed mixture with sifted dry ingredients. Pour into well greased and floured 9-inch square pan. Bake at 325 degrees for 45-50 minutes.

Mrs. J. S., Brookings Co., S. D.

RAISIN CAKE

1 c raisins
2½ c hot water
½ c lard
2 eggs
2 c sugar
3 c flour
½ tsp. salt
1 tsp. cinnamon
1 tsp. cloves
1 tsp. nutmeg
1 tsp. ginger
1 tsp. allspice
2 tbsps. cocoa
2 tsps. soda

Steam raisins in hot water. Cream lard, eggs and sugar. Sift together remaining ingredients. Drain raisins, reserving 1 cup water. Add water to creamed mixture alternately with dry ingredients; mix well. Add raisins. Bake at 350 degrees for 30-40 minutes.

Mrs. P. T., Stutsman Co., N. D.

CAKES/Spice

HONEY SPICE CAKE

- 2½ c flour
- ½ tsp. salt
- 1 tsp. soda
- 1 tsp. baking powder
- 1 tsp. cinnamon
- ½ tsp. ginger
- ¼ tsp. allspice
- ½ c butter or shortening
- 1 c honey
- 1 egg
- 1 c milk

Sift together dry ingredients. Cream butter or shortening; add honey; beat well. Add egg; mix well. Add dry ingredients and milk alternately; mix well. Bake in greased pan at 350 degrees for 40 minutes. Serve with chocolate sauce or whipped cream

Mrs. N. T., Dunn Co., N. D.

SPUD AND SPICE CAKE

- 1¾ c sugar
- 1 c cold mashed potatoes
- ¾ c shortening
- 1 tsp. cinnamon
- ½ tsp. salt
- ½ tsp. nutmeg
- 3 eggs
- 1 tsp. soda
- 1 c buttermilk or sour milk
- 2 c flour
- ¾ c chopped nuts
- 2 tbsps. flour

Cream thoroughly sugar, potatoes, shortening, cinnamon, salt and nutmeg. Add eggs; mix thoroughly. Combine soda and buttermilk or sour milk; add alternately to creamed mixture with 2 cups flour, beginning and ending with flour. Coat chopped nuts with 2 tbsps. flour; stir into batter. Pour into lightly greased and floured pan. Bake at 350 degrees for 50-60 minutes.

Mrs. W. L., Stark Co., N. D.

CAKES/Sponge

HOT MILK SPONGE CAKE

- ¾ c milk
- 1½ tbsps. butter
- 3 eggs
- 1½ c sugar
- ½ tsp. salt
- 1½ tsps. vanilla
- 1½ c sifted cake flour
- 1½ tsps. baking powder

Heat milk in double boiler; add butter. Beat eggs until thick and foamy. Gradually add sugar to eggs, beating constantly until well blended. Add salt and vanilla; beat 4 minutes. Sift together dry ingredients; add to egg mixture. Stir gently until dry ingredients are just mixed. Stir hot milk mixture into batter. Pour batter into a 13x9x2-inch pan lined with waxed paper. Bake at 350 degrees for 30 minutes.

Mrs. R. H., Hanson Co., S. D.

OLD-FASHIONED JELLY ROLL

¾ tsp. baking powder
¼ tsp. salt
4 eggs
¾ c sugar
¾ c sifted cake flour
1 tsp. vanilla
confectioners sugar
jelly

Mix together baking powder, salt and eggs. Place over a bowl of hot water; beat, gradually adding sugar. Beat until mixture becomes thick and light colored. Remove from hot water; fold in flour; add vanilla. Turn into a 15x10-inch pan lined with greased paper. Bake at 400 degrees for 13 minutes. Quickly cut off crisp edges of cake; turn from pan at once onto a cloth sprinkled with confectioners sugar. Remove paper. Spread with jelly almost to edge of cake. Roll quickly. Wrap in cloth and cool on rack.

Mrs. W. R., McIntosh Co., N. D.

White & Yellow/CAKES

GOLD ANGEL CAKE

12 egg yolks
¾ c boiling water
2 tsps. grated orange rind
1 tbsp. fresh orange juice
1¾ c cake flour
1¼ c sugar
2 tsps. baking powder
¼ tsp. salt

Beat egg yolks until thick and lemon colored. Add boiling water and continue beating five minutes. Add orange rind and juice. Sift together remaining ingredients; fold into egg yolk mixture. Bake in an ungreased 10-inch angel food tube pan at 325 degrees for 1 hour.

Mrs. A. Z., McIntosh Co., N. D.

MARTHA WASHINGTON CAKE

1 c shortening
2 c sugar
4 egg yolks
1 tsp. vanilla
3 c flour
¼ tsp. salt
3 tsps. baking powder
1 c milk
4 egg whites, stiffly beaten

Cream shortening and sugar. Add egg yolks and vanilla; beat well. Add sifted dry ingredients alternately with milk. Fold in egg whites. Bake in 3 waxed paper lined 8-inch round layer pans. Bake at 350 degrees for 30 minutes. Ice with your favorite chocolate frosting.

Mrs. R. M., Lake Co., S. D.

DELICIOUS FROSTING

2 c sugar
½ c milk
½ c shortening
1 c chocolate, caramel or butterscotch chips

Bring to boil sugar, milk and shortening. Remove from heat; add chips. Mix well as they melt. Beat until spreading consistency.

Mrs. O. R., Spink Co., S. D.

MARSHMALLOW FROSTING

12 marshmallows
3 tbsps. water
2 tbsps. butter
1 sq. chocolate
confectioners sugar
½ tsp. vanilla

Melt marshmallows, water, butter and chocolate in double boiler over low heat. Add enough confectioners sugar to achieve spreading consistency. Add vanilla.

Mrs. E. S., Barnes Co., N. D.

CARAMEL FROSTING

¼ c butter or margarine
1/3 c cream
¾ c brown sugar
½ tsp. vanilla
1¾ c sifted confectioners sugar

Melt butter; blend in cream and brown sugar. Boil for 1 minute; cool slightly. Add vanilla and confectioners sugar; beat until smooth.

Mrs. J. K., Nelson Co., N. D.

MAPLE FROSTING

¾ c brown sugar
2½ tbsps. water
1 egg white
¼ tsp. maple flavoring

Cook brown sugar, water and egg white in double boiler; beating constantly until frosting forms peaks, about 7 minutes. Remove from heat; add maple flavoring.

Mrs. G. W., Carter Co., Mont.

GREEN MINT FROSTING

1 c sifted confectioners sugar
1 tsp. butter
1 1/3 tbsps. milk
⅛ tsp. peppermint extract
green food coloring

Mix until smooth the confectioners sugar, butter and milk. Add extract and food coloring; mix well. Covers 8-inch square cake.

Mrs. C. H., Clay Co., Minn.

COOKIES & CANDY

Just about every family has a favorite cookie that Mom bakes again and again. The recipes on these pages are favorites that have kept cookie jars filled in many Dakota homes. Many of them were sent to THE DAKOTA FARMER with comments like: "I have to bake a batch of these every week. The children love them!"

BANANA BARS

- 1 c sifted flour
- ¾ c sugar
- ½ tsp. baking powder
- ½ tsp. salt
- ¼ tsp. soda
- ¾ tsp. cinnamon
- ¼ tsp. cloves
- ¼ tsp. allspice
- ¼ c butter
- ½ c mashed bananas
- 1 egg
- ¼ c milk
- 1/3 c chopped pecans (optional)

Sift together dry ingredients. Cream butter; add bananas; cream thoroughly. Add egg; beat well. Add dry ingredients alternately with milk. Add pecans, if desired. Spread in well-greased and floured 13x9-inch pan. Bake at 350 degrees for 22-25 minutes.

Mrs. L. R. N., Ransom Co., N. D.

CHERRY FUDGE BARS

- 1 (4-oz.) pkg. chocolate pudding and pie filling
- 1 pkg. chocolate cake mix
- 1 c chopped maraschino cherries
- 1 c chopped nuts
- 1 (6-oz.) pkg. semi-sweet
- 1 chocolate chips

Cook pudding mix as directed on package. Blend dry cake mix thoroughly into hot pudding. Add cherries. Pour into greased and floured 15x10x1-inch pan. Sprinkle with nuts and chips. Bake at 350 degrees for 25-30 minutes.

Mrs. E. W. Z., Dickey Co., N. D.

CHOCOLATE BROWNIES

- 1 pkg. yellow cake mix
- 1 pkg. instant chocolate pudding
- 2 eggs
- 2 c milk
- 1 c chopped nutmeats

Frosting:

- 3 egg yolks
- 1 c evaporated milk
- 1 c sugar
- ½ c butter or margarine
- 1 tsp. vanilla
- 1 c chopped pecans or black walnuts
- 1½ c coconut

Beat together cake mix, pudding, eggs and milk. Stir in nuts. Pour into greased and floured 10x15x1-inch jelly roll pan. Bake at 350 degrees for 40 minutes. Cool; frost.

Frosting: Beat together egg yolks and small amount of evaporated milk. Combine egg yolk mixture, remaining milk, sugar and butter or margarine; cook over medium heat, stirring constantly, until thick (about 12 minutes). Add vanilla, nuts and coconut. Beat to spreading consistency.

Mrs. W. H., Brookings Co., S. D.

ONE-BOWL BROWNIES

- 2½ sqs. chocolate
- 1 c shortening or lard
- 4 eggs
- 1 c brown sugar
- 1 c sugar
- 1 tsp. vanilla
- 1 c flour
- 1 tsp. baking powder
- ½ tsp. salt
- 1 c broken walnuts

Frosting:

- 2 sqs. baking chocolate
- 2 tbsps. butter
- ½ c milk
- 1½ c sifted confectioners sugar
- 1 tsp. vanilla

Melt together chocolate and shortening or lard; cool. Add eggs; beat well. Add remaining ingredients. Mix well; pour into greased and floured pan. Bake at 350 degrees for 40-45 minutes. Frost while slightly warm.

Frosting: Melt together chocolate and butter. Add milk; stir until thickened. Remove from heat; add confectioners sugar and vanilla; mix well. Cold milk or additional confectioners sugar may be added to achieve proper spreading consistency.

Mrs. M. D., McIntosh Co., N. D.

RHUBARB BARS

Crust:

- 1 c flour
- 5 tbsps. confectioners
- ½ c butter or margarine

Filling:

- 2 eggs, beaten
- 1½ c sugar
- ¼ c flour
- 2 c cut-up rhubarb

Combine crust ingredients; pat in 7½x11-inch pan. Bake at 350 degrees for 15 minutes.

Filling: Mix together filling ingredients; spread on crust. Bake at 350 degrees for 35 minutes.

Mrs. T. McP., Grant Co., S. D.

SPECIAL K BARS

- 1 c white syrup
- 1 c sugar
- 1½ c creamy or chunk style peanut butter
- 6 c Special K cereal
- 1 (6-oz.) pkg. chocolate chips
- 1 (6-oz.) pkg. butterscotch chips

Combine syrup and sugar; bring to a full boil and remove from heat. Add peanut butter; stir until dissolved. Add cereal; stir until completely coated with peanut butter mixture. Press mixture into a 9x13-inch greased pan. Melt together in double boiler the chocolate chips and butterscotch chips. Spread over bars and chill.

Mrs. R. K., McPherson Co., S. D.

7-LAYER BARS

¼ c butter
1 c graham cracker crumbs
1 c shredded coconut
1 (6-oz.) pkg. semi-sweet chocolate pieces
1 (6-oz.) pkg. butterscotch pieces
1 (15-oz.) can sweetened condensed milk
1 c chopped nuts

Melt butter in 13x9x2-inch pan; layer the remaining ingredients in the order given. Bake at 350 degrees for 30 minutes. Cool and cut into bars.

Ms. M. G., Custer Co., S. D.

LEMON DREAM BARS

½ c butter
1 c flour
¼ c confectioners sugar

Filling:
2 eggs
1 c sugar
2 tbsps. lemon juice
1/3 c coconut
lemon or coconut flavoring, optional
¼ c flour
½ tsp. baking powder
pinch of salt

Blend first three ingredients and pat into pan. Bake at 350 degrees for 8 minutes.

Filling: Beat together eggs, sugar, lemon juice and coconut. Add flavoring. Mix in flour, baking powder and salt. Spread over first layer and bake at 350 degrees for 25 minutes. Frost with confectioners sugar icing which has been flavored with lemon juice and lemon rind. Cut into bars.

Mrs. S. G., Minnehaha Co., S. D.

MINNESOTA HARVEST BARS

¼ c shortening, melted
1 c brown sugar
2/3 c cooked or canned pumpkin
2 eggs
½ tsp. vanilla
½ c sifted flour
½ tsp. baking powder
¼ tsp. soda
½ tsp. salt
½ tsp. cinnamon
½ tsp. nutmeg
½ tsp. ginger
½ c chopped dates
½ c walnuts
2 tbsps. flour
confectioners sugar

Combine melted shortening and brown sugar; stir until blended. Add pumpkin, eggs and vanilla. Combine dry ingredients; add to pumpkin mixture. Stir in dates and walnuts which have been combined with 2 tablespoons flour. Spread in greased 13x9x2-inch pan. Bake at 350 degrees for 30-35 minutes. While warm, sift confectioners sugar on top.

Mrs. J. J., Stark Co., N. D.

ROCKS

½ c shortening
1 c brown sugar
2 eggs, beaten
7/8 c flour
½ tsp. soda
1 tsp. cinnamon
1 tsp. salt
2 c oatmeal
1 cup raisins
½ c chopped dates
½ c chopped nuts

Cream together shortening and brown sugar; add eggs. Sift together flour, soda, cinnamon and salt; add to creamed mixture. Add remaining ingredients. Drop by spoonfuls onto buttered baking sheet. Bake at 375-400 degrees for about 20 minutes.

Mrs. J. P., Haakon Co., S. D.

PUMPKIN COOKIES

2 c brown sugar
2 c cooked or canned pumpkin
1 c salad oil
2 tsps. vanilla
4 c flour
2 tsps. soda
2 tsps. baking powder
1 tsp. salt
1 tsp. cinnamon
1 tsp. nutmeg
½ tsp. ginger
2 c raisins
1 c chopped nuts

Mix together sugar, pumpkin, oil and vanilla. Sift together dry ingredients; add to pumpkin mixture; stir until smooth. Blend in raisins and nuts. Drop by spoonfuls on greased baking sheet. Bake at 350 degrees for 12-15 minutes.

Mrs. J. D. A., Marshall Co., S. D

CHERRY ALMOND MACAROONS

1 (5-oz.) can slivered, blanched almonds (about 1¼ c)
¾ c sugar
3 egg whites
¼ c chopped maraschino cherries, drained
red food coloring (optional)

Grind dry, crisp almonds through finest blade of food grinder or chop as fine as possible in food blender. (Do not chop nuts by hand.) Combine nuts, sugar and egg whites in saucepan. Cook over medium heat, stirring constantly, 8-10 minutes or until path stays clean when spoon is drawn through. Remove from heat; stir in cherries and, if desired, coloring. Drop level tablespoonfuls of dough on two greased and lightly floured baking sheets. Let stand at room temperature until cool. Bake at 300 degrees for 20 minutes or until delicately golden brown. Remove from sheets immediately.

Mrs. D. B., Moody Co., S. D.

PEANUT BUTTER CRISSCROSSES

- 1 c shortening
- 1 c sugar
- 1 c brown sugar
- 2 eggs
- 1 tsp. vanilla
- 1 c peanut butter
- 3 c sifted flour
- 2 tsps. soda
- ½ tsp. salt

Cream together shortening, sugars, eggs and vanilla. Stir in peanut butter. Sift together dry ingredients; stir into creamed mixture. Drop by rounded teaspoonfuls on ungreased cookie sheet. Press with back of floured fork to make crisscross. Bake at 350 degrees for about 10 minutes.

Mrs. P. K., Polk Co., Minn.

ORANGE COOKIES

- 2 c sugar
- 1 c shortening
- 3 eggs
- 1 c buttermilk
- grated rind & juice of 2 oranges
- 1 tsp. soda
- 1 tsp. baking powder
- 3½ c flour

Cream together sugar and shortening. Beat in eggs; add buttermilk, orange juice and rind. Add dry ingredients. Drop by teaspoonfuls on greased cookie sheet. Bake at 400 degrees for 12-15 minutes. Frost with confectioners sugar frosting to which orange rind and orange juice have been added.

Note: More flour may need to be added, depending on the size of the oranges.

Mrs. N. L., Day Co., S. D.

BROWNIE DROPS

- ½ c shortening
- 1½ sqs. unsweetened chocolate
- 1 c brown sugar
- ½ c buttermilk or sour milk
- 1 egg
- 1 tsp. vanilla
- 1½ c sifted flour
- ½ tsp. baking powder
- ½ tsp. soda
- ¼ tsp. salt
- 1 (6-oz.) pkg. chocolate chips

Chocolate Glaze:

- ½ sq. unsweetened chocolate
- 2 tbsps. butter
- 1 c sifted confectioners sugar
- 1 tbsp. hot water
- ½ tsp. vanilla

Melt together shortening and chocolate squares. Add brown sugar, buttermilk or sour milk, egg and vanilla. Blend well. Sift together dry ingredients; add to chocolate mixture. Stir in chocolate chips. Drop by teaspoonfuls on greased baking sheet. Bake at 350 degrees for 10-12 minutes. While warm, spread with glaze.

Glaze: Melt together chocolate and butter. Blend in remaining ingredients.

Mrs. O. S., Ward Co., N. D.

CARROT COOKIES

¾ c butter
1 c sugar
1 c cooked, mashed carrots
1 egg, beaten
1¾ c flour
1 tsp. baking powder
2 tsps. vanilla or lemon juice (or 1 each)
pinch of salt

Frosting:
juice of 1 orange, strained
2 tbsps. butter, melted
1½ c confectioners sugar

Combine cookie ingredients in order given. Drop by spoonfuls on lightly greased cookie sheet. Bake at 375 degrees for 12-15 minutes. Frost.

Frosting: Combine frosting ingredients; beat well.

Mrs. F. H., Brown Co., S. D.

CHOCOLATE SUNDAE COOKIES

1 (6-oz.) pkg. chocolate chips melted
½ c shortening
1/3 c brown sugar
1 egg
1½ c sifted flour
½ tsp. soda
½ tsp. salt
¼ c maraschino cherry juice
2 tbsps. milk
½ c chopped nuts
¼ c chopped maraschino cherries
18 marshmallows
chocolate frosting
walnuts

Cream together shortening and brown sugar; blend in egg. Sift together flour, soda and salt; gradually add to creamed mixtuure with cherry juice and milk; mix well. Add nuts, cherries and melted chocolate. Drop by teaspoonfuls on greased cookie sheet. Bake at 350 degrees for 12 minutes. Cut marshmallows in half and place cut side down on hot cookies; cool. Frost with your favorite chocolate frosting; top with a walnut half.

Mrs. A. F., Stark Co., N. D.

MERINGUE COOKIES

3 egg whites
1 c sugar
¼ tsp. salt
1 tsp. vanilla
¾ c chopped nuts
¾ c chopped dates

Blend together egg whites, sugar, salt and vanilla in top of double boiler. Place over boiling water; beat with rotary beater until mixture stands in stiff peaks, scraping bottom and sides of pan occasionally with scraper. (Meringue will become lumpy if you don't scrape.) Immediately stir in walnuts and dates. Drop heaping teaspoonfuls of meringue immediately on two lightly greased baking sheets. Bake one sheet at a time at 300 degrees for 12-15 minutes, or until lightly browned. Remove from baking sheets immediately.

Variation: Candied cherries and coconut may be substituted for dates and nuts.

Mrs. D. B., Moody Co., S. D.

SOUR CREAM NUT COOKIES

1/3 c shortening
2/3 c sugar
1 egg
1 2/3 c flour
1 tsp. baking powder
¼ tsp. soda
½ tsp. salt
½ tsp. nutmeg
½ c sour cream
½ c chopped nuts
1 tbsp. sugar
¼ tsp. nutmeg

Cream together shortening, sugar and egg. Sift together flour, baking powder, soda, salt and nutmeg. Stir into creamed mixture alternately with sour cream. Add nuts; mix well. Drop dough by teaspoonfuls onto greased baking sheet. Flatten with greased bottom of glass dipped in sugar. Mix together sugar and nutmeg; sprinkle on top. Bake at 400 degrees for 8-10 minutes.

Miss P. D. A., Meade Co., S. D.

CREAM CHEESE COOKIES

¼ c butter or margarine
1 (8-oz.) pkg. cream cheese
1 egg yolk
¼ tsp. vanilla
1 pkg. yellow cake mix

Cream together butter and cheese. Blend in egg yolk and vanilla. Add cake mix in thirds, blending thoroughly. Chill dough 30 minutes. Drop dough by scant teaspoonfuls onto ungreased cookie sheet. Bake at 375 degrees.

Note: This dough is also suitable for use in a cookie press.

Mrs. E. W. Z., Dickey Co., N. D.

Filled/COOKIES

LEMON PUFFS

¾ c plus 2 tbsps. cake flour
¾ c cornstarch
2 tsps. baking powder
½ c butter
½ c sugar
2 eggs, well-beaten

Filling:
1 egg, lightly beaten
3 tbsps. lemon juice
rind of 1 lemon, grated
2/3 c sugar
1½ tbsps. butter, melted

Sift together flour, cornstarch and baking powder. Cream together butter and sugar; add eggs. Add flour mixture to creamed mixture; mix well Refrigerate 1 hour or more. Remove from refrigerator and shape dough into marble-sized balls. Place on greased cookie sheet and bake at 400 degrees for 6-8 minutes. The cookies should be uniform in size and yellow. Put 2 cookies together with filling; roll in confectioners sugar.

Filling: Mix together filling ingredients; cook in double boiler until thick.

Mrs. D. S., Grand Forks Co., N. D.

CHERRY CHOCOLATE COOKIES

1 c butter
½ c brown sugar
1 egg
2 c sifted flour
1 tsp. vanilla
maraschino cherries
sweet chocolate, melted
walnuts

Cream together butter and sugar. Add egg, vanilla and flour. Chill. Form dough around cherries and bake at 350 degrees until light brown. Cool. Dip cookies in chocolate. Roll each cookie in chopped walnuts. Set until firm.

Mrs. B. B., McIntosh Co., N. D.

RAISIN-FILLED COOKIES

4 c flour
1 tsp. soda
½ tsp. salt
1½ c sugar
1½ c shortening
½ c sour milk
2 eggs, beaten
1 tsp. vanilla

Filling
2 c ground raisins
1½ c water
¾ c sugar
2 tbsps. flour
1 tsp. vanilla
1 tsp. vinegar

Sift together dry ingredients. Cut in shortening as for pie crust. Combine sour milk, eggs and vanilla; add to dry ingredients; mix well. Chill dough thoroughly. Roll out and cut with cutter. Fill with filling; cover with another cookie; bake.

Filling: Combine raisins, water, sugar and flour; cook until thick. Remove from heat; add vanilla and vinegar. Cool.

Note: Keep cookies in well-covered container. They freeze very well.

Mrs. L. T., Lawrence Co., S. D.

CHOCOLATE SANDWICH DREAMS

½ c butter
1 c sugar
3 sqs. unsweetened chocolate, melted
2 tbsps. cream
1 tsp. vanilla
1½ c flour
½ tsp. baking soda
apricot jam

Chocolate Glaze:
1/3 c semi-sweet chocolate pieces
2 tbsps. light corn syrup
1 tsp. water

Cream butter. Gradually add sugar, chocolate, cream and vanilla. Sift together dry ingredients; add to creamed mixture; mix thoroughly. Roll into balls. Place on ungreased baking sheet; flatten with the bottom of a glass dipped in flour. Bake at 350 degrees for 9-12 minutes. Cool and place flat side of two cookies together with apricot jam sandwich style. Frost with chocolate glaze. Let stand until glaze hardens.

Chocolate Glaze: Melt chocolate pieces over boiling water. Add corn syrup and water; stir until smooth.

Mrs. H. B., Mercer Co., N. D.

LEMON COOKIES

1 c lard
1 c butter
1 c sugar
1 c confectioners sugar
½ tsp. salt
2 tsps. lemon extract
2 eggs
4 c flour
1 tsp. soda
1 tsp. cream of tartar

Cream together shortenings, sugar, salt and lemon extract. Add eggs and mix well. Sift together flour, soda and cream of tartar; add to creamed mixture, mixing well. Roll dough into walnut-sized balls and flatten with a glass dipped in sugar. Bake at 375 degrees for 10 minutes. Do not overbake cookies; they stay white in color.

Mrs. P. M. J., Grand Forks Co., N. D.

CHOCOLATE CRINKLES

½ c shortening
1 2/3 c sugar
2 tsps. vanilla
2 eggs
2 (1-oz.) sqs. chocolate, melted
2 c sifted flour
2 tsps. baking powder
½ tsp. salt
1/3 c milk
½ c nuts

Cream shortening, sugar and vanilla thoroughly. Beat in eggs, then chocolate. Sift together dry ingredients. Add to creamed mixture alternately with milk. Blend well after each addition. Stir in nuts. Chill 2-3 hours. Form in 1-inch balls; roll in confectioners sugar. Bake at 350 degrees 10-12 minutes.

Mrs. A. A., Spink Co., S. D.

WALNUT BUTTERBALLS

½ c butter or margarine
2 tbsps. honey
1 c sifted flour
¼ tsp. salt
1 tsp. vanilla
1 c finely chopped walnuts
confectioners sugar

Cream together butter or margarine and honey. Add flour and salt; mix well. Add vanilla. Stir in nuts; mix thoroughly. Chill dough for 1 hour. Form into walnut-size balls and place on greased baking sheet. Bake at 300 degrees for 35 minutes or until delicately brown. Roll cookies in confectioners sugar while still warm; reroll when cool.

Mrs. D. R., Day Co., S. D.

VERNA'S OATMEAL SUGAR COOKIES

¾ c butter
¾ c brown sugar
¾ c sugar
2 eggs
1 tsp. vanilla
1 tsp. salt
1½ c flour
1 tsp. soda
1 tsp. cinnamon
½ c raisins
3 c oatmeal
sugar

Cream together butter and sugars. Add eggs; add vanilla and salt. Sift together flour, soda and cinnamon. Stir raisins into flour mixture; add to creamed mixture. Stir in oatmeal. Roll into balls; dip in sugar; flatten. Bake at 350 degrees for 12-15 minutes.

Mrs. G. P., Bon Homme Co., S. D.

FORK COOKIES

1 c lard or shortening
1 cup sugar
1 c brown sugar
3 eggs, beaten
1 tsp. vanilla
3½ c flour
2 tsps. soda
1 tsp. cream of tartar
½ tsp. salt

Cream shortening. Cream in sugars; add eggs and vanilla. Sift together dry ingredients; add to creamed mixture. Roll into balls and place on greased cookie sheet; flatten with fork. Bake at 350 degrees for 15-20 minutes.

Note: This is a good basic cookie recipe. Add chocolate chips or 1 cup peanut butter for variation.

Mrs. O. R., Spink Co., S. D.

ANGEL FOOD COOKIES

½ c butter
½ c shortening
½ c brown sugar
½ c sugar
1 egg
2 c flour
1 tsp. cream of tartar
½ tsp. salt
1 tsp. soda
1 tsp. vanilla
walnuts
chocolate chips
sugar

Cream together butter, shortening and sugars. Add egg. Combine flour, cream of tartar, salt and soda; add to creamed mixture. Add vanilla. Add nuts and chocolate chips to suit taste. Roll dough into walnut-sized balls . Dip in water, then in sugar. Place far apart on greased cookie sheet. Bake at 425 degrees until done.

Mrs. J. S., Bon Homme Co., S. D.

SNOW BALLS

1 c soft butter
¾ c confectioners sugar
2 c cake flour
1 tsp. vanilla
1 c chopped nuts
confectioners sugar

Cream butter. Sift together confectioners sugar and flour; add to creamed butter. Add vanilla; stir in nuts. Roll into walnut-sized balls and bake at 300 degrees for 12-15 minutes. Roll in confectioners sugar while warm and again when cold.

Mrs. P. P., Cass Co., N. D.

GINGER SNAPS

¾ c shortening
1 c sugar
1 egg
4 tbsps. molasses
2 c flour
2 tsps. soda
1 tsp. baking powder
½ tsp. salt
1 tsp. ginger
1 tsp. cinnamon
¼ tsp. cloves

Cream together shortening and sugar; add egg and molasses; mix well. Sift together dry ingredients; add to creamed mixture. Mix ingredients well and place in refrigerator to chill. Roll dough into walnut-sized balls and roll in sugar. Bake at 325 degrees for 10-15 minutes or until brown. Do not overbake.

Mrs. H. E. M., LaMoure Co., N. D.

BUTTER CHIPS

1 c butter
1 c confectioners sugar
1 egg yolk
1 tsp. vanilla or almond extract
2 c flour
½ c walnuts or pecans

Mix ingredients in order given. Form in balls; flatten and bake at 350 degrees until done.

Mrs. D. G., Rollette Co., N. D.

Refrigerator/COOKIES

ANISE REFRIGERATOR COOKIES

1 c butter or margarine
2 c brown sugar
2 eggs, well-beaten
1 tbsp. crushed anise seed
1 c chopped pecans
3½ c flour
2½ tsps. baking powder
¼ tsp. salt

Cream together shortening and sugar; add beaten eggs. Add anise seed and pecans. Sift together dry ingredients; add to creamed mixture. Shape into rolls, wrap in waxed paper and chill thoroughly. Slice and bake at 350 degrees for 10-12 minutes.

Mrs. R. L., Washabaugh Co., S. D.

COCONUT REFRIGERATOR COOKIES

- 1½ c shortening
- 3 c brown sugar
- 2 eggs, well-beaten
- 2 tsps. vanilla
- 6 c flour
- 4 tsps. baking powder
- ½ tsp. salt
- 2 c shredded coconut

Cream together shortening and sugar. Add eggs and flavoring; mix well. Sift together flour, baking powder and salt; add to creamed mixture. Mix well. Add coconut; mix well. Divide dough in 6 parts, shaping each piece into a roll. Chill. When firm, slice very thin and bake at 375 degrees for 10-12 minutes.

Mrs. N. E. O., Pennington Co., Minn.

REFRIGERATED DATE-FILLED COOKIES

- 1 c shortening (part butter)
- 2 c brown sugar
- 2 eggs
- 1 tsp. vanilla
- 3½ c sifted flour
- 1 tsp. soda
- ½ tsp. salt

Filling:

- 1 c water
- 1 c sugar
- 1 tbsp. butter
- 2 c chopped dates
- ½ c chopped nuts

Cream together shortening, sugar, eggs and vanilla. Sift together dry ingredients; add to creamed mixture. Shape into 2½-inch rolls. Wrap in waxed paper and chill for a few hours or overnight. With sharp knife, slice cookie dough in 1/8-inch slices. Put 1 teaspoon filling on each cookie; top with another cookie; seal edges. Bake at 350 degrees for 8-10 minutes.

Filling: Combine all filling ingredients; cook until slightly thickened; cool.

Mrs. F. B., Burleigh Co., N. D.

FRUIT COOKIES

- 2 c brown sugar
- 1 c butter
- 1 c shortening
- 3 eggs, beaten
- 2 tsps. soda
- 2 tbsps. hot water
- 5-6 c flour
- 1 tsp. salt
- 1 tsp. nutmeg
- 1 tsp. cinnamon
- 1 c chopped nuts
- 1 bottle candied cherries
- 1 pkg. dates, cut-up
- 4 tsps. vanilla

Cream together sugar, butter and shortening. Dissolve soda in hot water. Add eggs and soda. Sift together flour, salt, nutmeg and cinnamon. Add to creamed mixture with nuts, cherries, dates and vanilla. Chill dough, then shape into rolls that can be sliced. When ready to bake, slice cookies and place on baking sheets. Bake at 400 degrees for 12 minutes or until golden brown.

Mrs. W. H. , Brookings Co., S. D.

PIN WHEELS

½ c butter
½ c sugar
¼ tsp. salt
1 tsp. vanilla
1 egg yolk
1½ c flour
½ tsp. baking powder
3 tbsps. milk
1 sq. chocolate or cake coloring

Cream together shortening and sugar; add salt, vanilla and egg yolk. Beat thoroughly. Add flour sifted with baking powder; add milk. Blend well. Divide dough into two portions; add melted chocolate or coloring to one-half. Roll out both pieces of dough. Make each the same size, shape and thickness. Place one sheet of dough on top of the other; roll up in jelly roll fashion. Wrap the roll in waxed paper and place in refrigerator overnight. Cut in slices ¼-inch thick. Place on cookie sheet and bake.

Variation: Instead of the chocolate or coloring, use cooked dates or apricots.

Mrs. L. G., Walsh Co., N. D.

Rolled/COOKIES

CRISP WHITE SUGAR COOKIES

4 c sifted flour
1 tsp. soda
1½ c sugar
½ tsp. salt
1½ c butter or shortening
½ c sour milk or buttermilk
2 eggs, slightly beaten
1 tsp. vanilla

Sift together flour, soda, sugar and salt; cut in shortening. Combine remaining ingredients; add to flour mixture; beat until smooth. Cover and chill overnight. Roll thin; cut with cookie cutter. Bake at 425 degrees for 8-10 minutes.

Note: These are ideal for decorating.

Mrs. R. H., McLean Co., N. D.

HONEY COOKIES

1 c sugar
1 c honey
1 tsp. vanilla
3 eggs, well-beaten
5 c flour
3 tsps. soda
1 tsp. salt

Combine sugar, honey and vanilla; mix well. Add eggs. Sift together flour and soda; add to honey mixture; mix well. Let stand overnight; roll out and cut with doughnut cutter. Bake on well-greased cookie sheet at 300 degrees for 10 minutes or until light brown. Freeze or store in cold place.

Mrs. K. E. P., Williams Co., N. D.

COOKIES/Rolled

AUNT SALLY COOKIES

1 c sugar
1 c shortening
2 eggs, well-beaten
½ c molasses
1 c sour milk
2 tsps. cream of tartar
4 tsps. soda
1 tbsp. ginger
1 tbsp. cinnamon
5 c flour
Icing:
1½ c sugar
½ c water
1 tsp. vinegar
2 egg whites, well-beaten
12 marshmallows, cut-up

Cream together sugar and shortening. Add eggs, molasses and sour milk; mix well. Sift together dry ingredients and add to creamed mixture; mix well. Chill dough 1 hour. Roll out on floured board; cut with cookie cutter. Bake at 350 degrees until done. Do not overbake. Ice.

Icing: Combine sugar, water and vinegar; boil to medium ball stage. Pour over beaten egg whites, adding marshmallows.

Mrs. W. W. M., Traill Co., N. D.

CHOCOLATE COOKIES

½ c sugar
½ c brown sugar
½ c shortening
2 eggs
6 tbsps. cocoa
1 tsp. vanilla
1 tsp. soda
2½-3 c flour
1 c sour milk

Cream together sugars and shortening. Add eggs, cocoa and vanilla; beat well. Sift together soda and flour; add alternately with sour cream. Chill 1 hour; roll out thin and cut with cookie cutter. Bake at 350 degrees until done. Frost, if desired. Cookies may be put together sandwich style with icing between.

Mrs. M. C., Brown Co., S. D.

COOKIES/Unbaked

NO-BAKE COCONUT SURPRISE

1 (3-oz.) pkg. cream cheese
1 tsp. chopped nuts
1 tsp. grated orange or lemon rind
¼ c flaked coconut, toasted

Beat cheese until light and fluffy. Add nuts and rind; roll in coconut.

Mrs. C. W., Renville Co., N.D.

NO-BAKE CHOCOLATE OATMEAL COOKIES

3 c oatmeal
6 tbsps. cocoa
1 c coconut
½ c chopped nuts
2 c sugar
½ c milk
¼ lb. margarine

Mix oatmeal, cocoa, coconut and nuts. Boil sugar, milk, nuts and margarine for 1½ minutes. Add to first mixture and mix well. Drop by teaspoonfuls on waxed paper.

Mrs. T. K., Brule Co., S. D.

NO-BAKE FUDGE COOKIES

2 c sugar
1 small can condensed milk
2 sqs. chocolate
1 tbsp. butter
1 tsp. vanilla
28 crushed graham crackers
1 (10-oz.) pkg. marshmallows, quartered

Bring sugar, milk and chocolate to boil; boil 3 minutes. Add butter; cool. Add remaining ingredients. Drop on waxed paper.

Mrs. C. W., Renville Co., N. D.

NO-BAKE PEANUT BUTTER COOKIES

½ c sugar
½ c white syrup
dash of salt
½ tsp. vanilla
2/3 c crunchy peanut butter
3 c corn flakes

Combine sugar, syrup, salt and vanilla. Boil until it bubbles around the edges. Remove from heat; add peanut butter; stir until melted. Pour over corn flakes; mix well. Drop on waxed paper; let cool. If mixture sets too quickly, place over bowl of hot water.

Mrs. P. K., Polk Co., Minn.

BOILED COOKIES

2 c sugar
½ c milk
¼ lb. margarine
4 tbsps. cocoa
2½ c quick oatmeal
¼ c nuts
2 tsps. vanilla
½ c peanut butter (optional)

Cook sugar, milk, margarine and cocoa 1½ minutes at a full boil. Add remaining ingredients; beat well. Spoon onto waxed paper.

Mrs. J. N. B., Moody Co., S. D.

BUTTERSCOTCH CLUSTERS

½ c peanut butter
2 pkgs. butterscotch chips
6 c corn flakes

Melt together peanut butter and chips. Pour over corn flakes; mix well. Drop by teaspoonfuls on cookie sheet.

Mrs. E. H., Burke Co., N. D.

NO-BAKE BARS

graham crackers
1 c brown sugar
1 c crushed graham crackers
½ c butter or margarine
1/3 c milk
½ c coconut and nuts
1 tsp. vanilla
1 egg, beaten

Line 8x10-inch pan with whole graham crackers. Combine remaining ingredients and boil until thick (about 5 minutes), stirring constantly. Pour into graham cracker lined pan. Top with whole graham crackers and frost with confectioners sugar frosting or nuts and coconut. Cool and cut into squares.

Mrs. L. N., Ward Co., N. D.

NO-BAKE APRICOT BITES

15 square graham crackers
¾ c butter
1 c sugar
½ c milk
1 egg, slightly beaten
1 c crushed graham crackers
1 c flaked coconut
1 c chopped nuts
¾ c finely chopped dried apricots
½ tsp. vanilla
15 square graham crackers

Glaze:
1 c sifted confectioners sugar
juice of 1 lemon
1 tbsp. butter, melted

Line 13½x18½-inch pan with graham crackers. Melt butter; stir in sugar, milk and beaten egg. Cook, stirring until mixture is slightly thickened (about 5 minutes). Add crushed graham crackers; mix well. Add coconut, nuts, apricots and vanilla. Spread over crackers in pan. Top with another layer of graham crackers.

Glaze: Combine glaze ingredients and glaze. Let stand until glaze sets; cut in bars.

Mrs. J. W. P., Lincoln Co., S. D.

NOODLE DROPS

1 pkg. chocolate chips
1 pkg. butterscotch chips
1 can chow mein noodles
1 can or small bag Spanish peanuts

Melt in double boiler chocolate and butterscotch chips. Add noodles and peanuts. Drop on waxed paper.

Mrs. E. H., Burke Co., N. D.

GRANDMA'S COOKIES

1 (1-lb.) pkg. marshmallows
1 (6-oz.) pkg. chocolate chips
4 tbsps. butter
1½ c peanuts
1½ c Rice Krispies

Melt in top of double boiler marshmallows, chocolate chips and butter. Stir until well blended; pour over remaining ingredients. Mix well and drop by teaspoonfuls on waxed paper.

Mrs. J. S., Ward Co., N. D.

Miscellaneous/COOKIES

RAISIN GRIDDLE COOKIES

3½ c flour
1 c sugar
1½ tsps. baking powder
1 tsp. salt
½ tsp. baking soda
1 tsp. nutmeg
1 c shortening
1 egg, beaten
½ c milk
1¼ c raisins

Sift together dry ingredients; cut in shortening until mixture is mealy. Combine egg and milk; blend. Add, with raisins, to dry mixture. Stir until ingredients are moistened and mixture holds together. Roll out on floured board to ¼-inch thickness; cut with 2-inch round cookie cutter. Heat griddle until moderately hot, or a few drops of water dance on it. Oil griddle lightly; place cookies on it. As the bottom of cookie browns, top will get puffy. Turn and brown other side. Serve hot from griddle.

Variation: Substitute 1 teaspoon grated lemon rind for raisins.

Note: To freeze, pack unbaked cookies, separated by foil, in freezer containers. Defrost 15-20 minutes; bake as directed.

Mrs. C. S., Meade Co., S. D.

CHOCOLATE TURTLES

½ c butter or margarine, melted
¾ c sugar
2 eggs, well-beaten
6 tbsps. cocoa or 1 sq. chocolate
1 c flour
1 tsp. vanilla

Combine all ingredients; mix well. Heat waffle iron and drop a teaspoon of dough on each section of iron. Bake 50-60 seconds. Lift off with fork.

Note: These cookies may be frosted with a chocolate confectioners sugar frosting.

Mrs. R. K., McCook Co., S. D.

ROSETTES

2 eggs, slightly beaten
1½ tsps. sugar
¼ tsp. salt
1 c milk
1 c sifted flour
red, green or white sugar

Mix together all ingredients except colored sugar; beat until smooth. Heat rosette iron; heat fat to 365 degrees. Dip iron into batter within ¼ inch from top. Fry in hot fat and drain well before removing rosette from the iron. Sprinkle with sugar.

Note: If it is difficult to remove rosettes from iron, they are not fried long enough. If fat blisters are present, the eggs were beaten too long. If batter drops from the iron, the fat isn't deep enough and iif rosettes aren't crisp, they were fried too fast.

Mrs. E. E., Bon Homme Co., S. D.

WELSH COOKIES

4 c flour
3 tsps. baking powder
¼ tsp. nutmeg
pinch of salt
2 c sugar
½ c butter
½ c shortening
1 c currants
3 eggs and enough milk to fill 1 cup

Sift flour, baking powder, nutmeg and salt; add sugar. Blend sugar and shortening into flour mixture. Add currants; add milk and eggs. Mix well. Roll out on floured board. Cut in rounds. Bake on ungreased griddle, turning to brown each side.

Mrs. C. T., Hand Co., S. D.

EGG YOLK PAINT FOR CUT-OUT COOKIES

Blend together 1 egg yolk and ¼ tsp. water. Divide mixture. Add food coloring. Place cookies on cookie sheet. Paint designs with small paint brush. Sprinkle with sugar. Bake as directed in recipe.

Note: If paint thickens while standing, add a few drops water.

Mrs. R. H., Day Co., S. D.

HOME CANDIED FRUIT

1 (No. 2) can sliced pineapple
1 c sugar
confectioners sugar
1 (8-oz.) jar maraschino cherries

Drain pineapple slices; place slices in electric fry pan set at 275 degrees. Sprinkle with sugar and cook slowly for 30 minutes or until mixture is just barely bubbling. Stir often and turn the pineapple occasionally. Let stand overnight in the syrup created from cooking. Place slices of fruit on cooling rack to dry — 2 days or more. Roll in confectioners sugar. Use remainder of syrup to cook cherries in the same manner. The two fruits can be cooked together but the pineapple will turn pink.

Miss P. D. A., Meade Co., S. D.

NO-COOK PEANUT CLUSTERS

1/3 c soft margarine or butter
1/3 c light or dark syrup
1 c cocoa
1 tsp. vanilla
3 tbsps. water
1 c nonfat dry milk
2 c sifted confectioners sugar
1 c salted peanuts

Blend margarine and syrup in large mixing bowl. Stir in cocoa; add vanilla and water. Blend in dry milk; then confectioners sugar. Mix well. Add peanuts. Drop from tip of spoon onto waxed paper. Allow clusters to remain uncovered for 2-3 hours to dry. Store between layers of waxed paper in covered container.

Mrs. A. H., Carter Co., Mont.

PARTY MINTS

1 (3-oz.) pkg. cream cheese
2 c sifted confectioners sugar
½ tsp. desired flavoring
food coloring

Cream one package of cream cheese; gradually add confectioners sugar, creaming well after each addition. Add flavoring and food coloring. If dough is not stiff enough, add more sugar. Form marble-sized balls, roll in granulated sugar and flatten or use mint molds. These freeze well.

Mrs. R. W. Foster Co., N. D.

FANNY FARMER FUDGE

4½ c sugar
1 (13-oz.) can evaporated milk
3 (6-oz.) pkgs. chocolate chips
¼ lb. butter
2 c walnuts
3 tsps. vanilla

Boil sugar and evaporated milk 10 minutes. Add chocolate chips, butter, walnuts and vanilla. Beat and pour into a greased pan.

Mrs. M. B., Brown Co., S. D.

CANDY

DIVINITY

3 c sugar
½ c light corn syrup
2/3 c water
2 egg whites
⅛ tsp. salt
½ tsp. vanilla
1 c nuts (optional)

Boil sugar, syrup and water to hard ball stage. Beat egg whites and salt at high speed in large mixer bowl for 3 minutes. Pour syrup slowly into beaten whites while beating at medium high speed. Continue beating until mixture passes the glossy stage. Add vanilla and continue beating until mixture forms a peak when beater is raised. Stir in chopped nuts. Drop candy by spoonfuls on waxed paper. (Instead of using chopped nuts, you may omit them and press a pecan or walnut half on the top of each piece.)

Mrs. H. K., Kingsbury Co., S. D.

CARAMEL WALNUT LOGS

½ c sweetened condensed milk
3 c confectioners sugar
½ tsp. maple flavoring
½ lb. (about 28) caramels
1 tbsp. butter
1 tbsp. water
1½ c chopped walnuts

Combine condensed milk, sugar and flavoring. Beat hard until smooth; then knead with hands for a few minutes. Divide mixture into four parts; shape each into a 5-inch long roll. Combine caramels, butter and water in top of double boiler and heat over boiling water until melted and smooth. Spread chopped walnuts on piece of waxed paper. Working with one candy roll at a time, coat one side of roll with caramel and drop onto walnuts; coat remainder of roll, completely covering it with walnuts. Repeat with remaining three candy rolls. Wrap each in plastic film and refrigerate until firm. Cut into ½-inch slices.

Mrs. L. H., Kittson Co., Minn.

PENUCHE PATTIES

3 c brown sugar
¼ c butter
1 c light cream
1 c chopped pecans
1 tsp. vanilla
pecan halves

Combine sugar, butter and cream in saucepan. Cook to 238 degrees (soft ball stage) over medium heat, stirring occasionally. Remove from heat and cool to 130 degrees. Add pecans and vanilla. Beat until mixture changes color and thickens. Pour quickly into a buttered 9x9-inch pan. Cut into 25 squares. Roll each square into a ball; flatten into patties on waxed paper and top with pecan.

Mrs. A. S., Bowman Co., N. D.

CHEWY CARAMELS

2 c dark corn syrup
1 c light cream
2 tbsps. butter
1 tsp. vanilla

Combine syrup, cream and butter. Bring mixture to a boil and continue cooking until it reaches the hard crack stage in cold water, or 300 degrees on a candy thermometer. Remove from heat; add vanilla. Pour into greased 8-inch square pan. Mark in squares when slightly cool; cut when cold. Fold squares over once for thicker caramels. Wrap in waxed paper.

Mrs. M. L., Butte Co., S. D.

CRACKERJACKS

1/3 c molasses
1 tsp. vinegar
1 c sugar
1/3 c boiling water
1 tbsp. butter
pinch of soda
1 tsp. vanilla
popped corn

Cook to hard ball stage molasses, vinegar, sugar and boiling water; add butter, soda and vanilla. Pour over popped corn.

Mrs. H. B., Barnes Co., N.D.

DATE NUT LOGS

1 tbsp. butter or margarine
¼ c light or dark syrup
½ tsp. vanilla
3 tbsps. nonfat dry milk
¼ tsp. salt
2 c confectioners sugar
1 c finely cut dates
1 c chopped nuts

Blend butter and syrup; stir in vanilla. Combine dry milk, salt and sugar; add to syrup mixture. Mix in chopped dates. Stir and knead until thoroughly mixed. Form into small rolls; dip each roll into chopped nuts.

Miss H. C. H., Pope Co., Minn.

ALMOND BARK

2/3 c milk, scalded
2 c sugar
1 tsp. white syrup
dash of salt
2 tbsps. butter
1 tsp. vanilla
1 c chopped almonds

Cook to soft ball stage milk, sugar and syrup; add salt. Remove from heat and add butter. Cool slightly. Add vanilla; beat until thick. Do not overcook or overbeat. Add almonds. Pour into buttered pan.

Miss M. L., McCook Co., S. D.

CANDY

ANISE CANDY

1 c sugar
1/3 c white syrup
1/3 c water
red food coloring
1 tsp. anise flavoring

Combine all of the above ingredients except food coloring and flavoring; boil until mixture crackles when dropped in cold water. Add coloring and flavoring. Pour mixture into butter plates; let cool long enough to handle. Cut into strips. This must be handled quickly as it hardens rapidly.

Mrs. D. M., Hyde Co., S. D.

GOLDEN FUDGE

3 c sugar
¼ c light corn syrup
3 tbsps. butter
½ tsp. salt
1 c evaporated milk
½ c water
2 tsps. vanilla
green and red candied cherries

Combine sugar, syrup, butter, salt, evaporated milk and water in medium-size heavy saucepan. Heat, stirring constantly, to boiling, then cook rapidly, stirring several times, to 238 degrees (soft ball stage). Remove from heat at once. Add vanilla but do not stir in. Cool in pan to 110 degrees (lukewarm). Beat 2-3 minutes or just until set; cut into squares. Decorate each with slivered cherries. Let stand until firm.

Mrs. G. S., Day Co., S. D.

CARAMEL APPLES

15-20 small apples
4 c sugar
1 c light corn syrup
2 2/3 c evaporated milk

Wash and dry apples thoroughly; insert wooden skewers. Combine sugar, syrup and 2/3 cup evaporated milk in 3-quart heavy saucepan; stir to blend well. Heat slowly until sugar dissolves, stirring constantly. Cook briskly to a thick syrup, stirring constantly. Slowly add remaining evaporated milk (keep mixture boiling briskly) and cook to firm ball stage (242 degrees). Stir constantly to prevent scorching. Remove from heat; let stand until candy stops bubbling. Work quickly, dipping apples one at a time into caramel syrup. Place on buttered aluminum foil or buttered baking sheet. If caramel coating becomes too hard for dipping, add a little evaporated milk and reheat, stirring to keep smooth. The caramel should be kept hot so coating will not be too heavy.

Mrs. C. H., Hettinger Co., N. D.

PEANUT BRITTLE

3 c sugar
1 c water
½ c white corn syrup
¼ c butter
1 lb. raw Spanish peanuts (with skins)
1 tsp. soda
1 tsp. vanilla

Boil sugar, water and corn syrup until it threads. Add butter and peanuts; cook slowly to 300 degrees, stirring occasionally. Add soda and vanilla. (It will ofam up.) Turn out onto 2 buttered baking sheets; pull out thin as it cools. When cold, break into pieces with knife handle.

Ms. H. N., Roosevelt Co., Mont.

BUTTER CRUNCH

1 c sugar
½ tsp. salt
¼ c water
½ c butter
½ c finely chopped walnuts
1 (6-oz.) pkg. chocolate chips
1 c finely chopped walnuts

Combine sugar, salt, water and butter. Heat to boiling and cook to light crack stage (285 degrees). Add walnuts and pour onto a well-greased cookie sheet. Cool. Spread half of melted chocolate pieces over top of candy. Sprinkle with half of the walnuts. Let stand until chocolate is set. Turn and spread with remaining half of chocolate and walnuts. Break into pieces to serve.

Mrs. M. I., Campbell Co., S. D.

DESSERTS

Every family has its sweet tooth . . . you know, the one who is the true authority on sweets and desserts. It's for the sweet tooth in your family that we have included this irrestible collection of Dakota desserts — baked, frozen, chilled and whipped. If you're craving something sweet, here's the place to look.

APPLE CRISP

¾ c butter or margarine
½ tsp. soda
⅛ tsp. salt
1 c brown sugar
1½ c flour
1 c quick-cooking oatmeal
2 c diced and peeled apples
½ c sugar
4 tbsps. flour
¼ tsp. cinnamon
¼ tsp. nutmeg

Combine butter, soda, salt, sugar, flour and oatmeal; mix well. Place half mixture in bottom of well-greased and floured 9-inch square pan. Combine remaining ingredients. Add apples to crumb crust mixture; sprinkle remaining mixture over apples; press down firmly. Bake at 350 degrees for 40-45 minutes, or until apples are soft. Serve hot or cold with whipped cream or ice cream.

Variation: Pie mixes, such as apple, cherry or blueberry, may be substituted for apple mixture.

Mrs. D. G., Brookings Co., S. D.

APPLE DUMPLINGS

apples
2 c flour
2 tsps. baking powder
1 tsp. salt
1 c butter
water
2 c sugar
1 c water

Peel apples; cut in half and remove core. Make pastry dough of flour, baking powder, salt, butter and enough water to make a soft dough. Roll out about ¼-inch thick. Cut in strips and roll around an apple (both halves). Place in large casserole or deep cake pan to bake. Combine sugar and water; boil. Pour over apples and bake until crust is brown.

Mrs. A. H., Stutsman Co., N. D.

BAKED RICE

3 eggs
½ c sugar
1 c cream
1 tsp. vanilla
pinch of salt
½ c cooked, drained rice
cinnamon
raisins (optional)

Beat together eggs, sugar, cream, vanilla and salt. Add rice. Place in casserole and sprinkle with cinnamon. Bake at 325 degrees for 45 minutes or until done.

Mrs. S. H., Emmons Co., N. D.

APPLE PUDDING

- 4 c peeled and diced apples
- 2 c sugar
- 2 c flour
- 2 tsps. soda
- 1 tsp. cinnamon
- ½-1 c nuts
- 2 eggs
- ½ tsp. salt

Combine apples and sugar; let set for ½ hour. Add remaining ingredients in order given; mix well. Bake at 350 degrees for 30 minutes or until done. Serve with whipped cream.

Mrs. W. R., Douglas Co., S. D.

CARAMEL TAPIOCA

- 5½ ozs. tapioca
- 1 qt. water
- 1 c brown sugar
- ¼ tsp. salt
- 1 tbsp. butter
- ¾ cup cream
- ½ tsp. vanilla

Soak tapioca overnight in water. In the morning add brown sugar and salt. Bake at 350 degrees for 2 hours. Stir after first hour. After second hour remove from oven and add butter, cream and vanilla; mix well. Serve hot or cold with whipped cream topping.

Mrs. W. W., Jerauld Co., S. D.

CARROT PUDDING

- 1½ c bread crumbs
- ¼ tsp. salt
- 1 tsp. baking powder
- 1 tbsp. shortening
- ½ c grated carrots
- 1 c molasses or brown sugar
- ½ c chopped dates
- ½ c raisins
- ½ c chopped figs
- ½ c ground nuts
- 2 tbsps. lemon juice
- 1 tsp. grated lemon rind

Raisin Sauce:

- ½ c raisins
- 1 c boiling water
- ¼ c chopped citron
- ¾ c sugar
- 1 tsp. cornstarch
- 1 tbsp. butter
- ½ tsp. lemon juice

Mix together bread crumbs, salt and baking powder. Add all remaining pudding ingredients; mix thoroughly. Fill a greased mold 2/3 full and steam 3-4 hours. Serve with lemon or raisin sauce.

Raisin Sauce: Combine raisins, boiling water and citron; simmer until raisins and citron are tender (about 1 hour). Sift together sugar and cornstarch; add to raisin mixture. Mix in butter and lemon juice. Continue cooking for 10 minutes.

Mrs. W. T., Walsh Co., N. D.

BREAD PUDDING

2 tbsps. butter
3 eggs, slightly beaten
½ c sugar
¾ tsp. salt
1 tsp. vanilla
1 qt. milk
3½ c dry bread cubes
raisins (optional)

Combine all of the above ingredients; bake at 350 degrees for 45-50 minutes. Serve warm or cold.

Mrs. K. K., Sheridan Co., Mont.

CHEESE CAKE

3 c flour
1 c butter
¼ c sugar
2 egg yolks
1 tbsp. grated lemon rind
lemon juice
¼ c cold water

Filling:
1½ lbs. cottage cheese, creamed fine or sieved
5 egg yolks, beaten
1½ c sugar
1 c milk
1 pt. cream
2 tbsps. cornstarch
1 tsp. vanilla
5 egg whites, beaten

Combine all dough ingredients; press into pan.

Filling: Combine filling ingredients; pour over dough. Bake at 350 degrees until filling is well set and dough is brown on bottom.

Ms. E. H., Griggs Co., N. D.

DEEP DISH APPLE PIE

8 c peeled and sliced tart apples
¾-1 c sugar
1 tsp. nutmeg
1 tbsp. butter
pastry for single-crust pie

Custard Sauce:
1 c cream
2 egg yolks, slightly beaten
¼ c sugar

Place apples in 9x9x2-inch baking dish. Top with sugar-nutmeg mixture. Dot with butter. Cover with pastry; slash crust to let steam escape. Bake at 425 degrees for 40 minutes. Serve with custard sauce.

Custard Sauce: Combine ingredients in double boiler; cook, stirring constantly, until mixture coats a spoon.

Mrs. W. H., Brown Co., S. D.

OZARK PUDDING

1 egg
¾ c sugar
1 tsp. vanilla
¼ c flour
1¼ tsps. baking powder
¼ tsp. salt
1 c peeled apples, coarsely chopped
½ c chopped nuts

Beat egg, sugar and vanilla until light and fluffy. Sift together dry ingredients; add to first mixture, blending by hand. Fold in apples and nuts. Pour into well-greased 9-inch square pan. Bake at 350 degrees for 25 minutes. Top with ice cream or whipped cream.

Note: Use either white sugar for a delicate flavor, or brown sugar for a caramel flavor. Pears or peaches may be substituted for apples.

Mrs. J. S., Ward Co., N. D.

PUFF PUDDING

¼ c butter
½ c sugar
1 tsp. grated lemon rind
2 egg yolks
3 tbsps. lemon juice
2 tbsps. flour
¼ c whole wheat flakes cereal
1 c milk
2 egg whites, stiffly beaten

Cream together butter, sugar and lemon rind. Add egg yolks; beat until light and fluffy. Blend in lemon juice, flour, cereal and milk. (Mixture will look curdled.) Fold in egg whites. Pour into greased custard cups or 1-quart casserole; place in pan of hot water. Bake at 325 degrees for about 40 minutes for cups or 1 hour and 15 minutes for casserole. When done, top will spring back when lightly touched. Pudding will have cake-like layer on top with custard below. Serve warm or cold.

Mrs. M. J. S., Charles Mix Co., S. D.

RHUBARB DEMIJOHN

2 c finely-chopped rhubarb
½ c sugar
sugar
Cream Batter:
1 egg
¾ c sugar
½ c sour cream
1 c flour
1 tsp. baking powder
¼ tsp. salt
½ c chopped nut meats

Mix together rhubarb and ½ cup sugar. Place rhubarb in buttered 1-quart baking dish; sprinkle with sugar. Top with cream batter. Bake at 375 degrees until done. Serve warm with plain or whipped cream.

Cream Batter: Combine egg, sugar, sour cream and flour; add remaining ingredients.

Mrs. R. H., Cass Co., N. D.

RHUBARB DESSERT

1¼ c flour
½ c shortening
2 tbsps. sugar
4 egg yolks
2 tbsps. flour
1½ c sugar
½ c evaporated milk
5 c chopped rhubarb
4 egg whites
6 tbsps. sugar

Crumble ¼ cup flour, shortening and 2 tablespoons sugar; press into 9x11-inch pan. Bake at 325 degrees for 20 minutes. Beat until white and thick egg yolks, 2 tablespoons flour and sugar; add milk and rhubarb. Spread over crust; bake for 40 minutes. Beat egg whites with 6 tablespoons sugar; pile on rhubarb mixture. Bake until meringue is brown. Serve the same day as baked, as bottom crust will soften if kept overnight.

Mrs. T. McP., Grant Co., S. D.

RHUBARB CRUNCH

3 c fresh chopped rhubarb
1 tbsp. water
3 tbsps. cornstarch
½ c sugar
1 pt. fresh strawberries, washed and hulled
red food coloring
1 c flour
½ tsp. soda
¼ tsp. salt
1 c oatmeal
¾ c brown sugar
½ c butter, melted

In medium-sized saucepan, cover and cook slowly over low heat rhubarb and water. Stir occasionally until rhubarb is of sauce consistency. Combine cornstarch and sugar. Add a small amount of cooked rhubarb; mix well. Add to rhubarb mixture in pan. Cook until thickened, about 10 minutes, stirring frequently. Stir in strawberries and food coloring. Sift together flour, soda and salt. Add remaining ingredients; mix thoroughly. Pat 2/3 of oat mixture evenly into an 8-inch pan. Spread rhubarb filling over oat mixture. Sprinkle remaining oat mixture over filling. Bake at 350 degrees for 30-35 minutes.

Miss F. G., Butte Co., S. D.

EASY DESSERT

1 pkg. white cake mix
1 c chopped rhubarb
brown sugar
nuts

Prepare cake mix according to package directions; stir in rhubarb. Pour into greased 9x13-inch cake pan. Sprinkle a mixture of brown sugar and nuts on top. Bake as directed. Serve with ice cream.

Mrs. J. S., Ward Co., N. D.

DESSERTS/Baked

STRAWBERRY SHORTCUT CAKE

- 1 c miniature marshmallows
- 2 (10-oz.) pkgs. frozen, sliced strawberries in syrup, thawed
- 1 (3-oz.) pkg. strawberry gelatin
- 1½ c sugar
- ½ c solid shortening
- 3 eggs
- 2¼ c flour
- 3 tsps. baking powder
- ½ tsp. salt
- 1 c milk
- 1 tsp. vanilla

Generously grease bottom only of 13x9-inch pan; sprinkle marshmallows evenly over bottom of pan. Combine thawed strawberries and syrup with dry gelatin; thoroughly mix and set aside. Combine remaining ingredients; blend at low speed until moistened. Beat at medium speed 3 minutes, scraping sides of bowl occasionally. Pour batter evenly over marshmallows in pan. Spoon strawberry mixture evenly over batter. Bake at 350 degrees for 45-50 minutes, or until done and golden brown. Serve warm or cool with ice cream or whipped cream.

Mrs. F. S., Sargent Co., N. D.

DESSERTS/Frozen

LEMON-ORANGE AND BANANA SHERBET

- 3 c sugar
- 3 c water
- juice of 3 lemons
- juice of 3 oranges
- 3 bananas, mashed
- 1 c whipped cream

Boil sugar and water for 3 minutes; let cool. Add fruit juices; stir in mashed bananas. Fold in whipped cream. Place in freezer. When mixture starts to freeze remove from freezer and mix well. Return to freezer until frozen.

Mrs. G. W., Dewey Co., S. D.

FROZEN PEACH CUSTARD

- 2½ c milk
- ¾ c sugar
- ¼ tsp. salt
- 1 tbsp. flour
- 3 eggs, beaten
- 1 tsp. vanilla
- 1½ c sliced peaches

Scald 2 cups milk in top of a double boiler. Combine sugar, salt and flour. Add ½ cup milk to make a smooth paste. Add flour paste to hot milk; cook until thickened, stirring constantly. Add beaten eggs; cook 2 minutes longer. Chill. Add vanilla and sliced peaches. Pour into freezing tray and freeze, stirring several times during freezing.

Miss C. A. M., Hand Co., S. D.

LIME-ORANGE DESSERT MOLD

2 envelopes unflavored gelatin
1/3 c sugar
⅛ tsp. salt
1½ c boiling water
1½ tsps. grated lime rind
½ c lime juice
1 (7-oz.) bottle lemon-lime carbonated beverage
1 c sour cream
1 (11-oz.) can mandarin orange sections, drained and cut into quarters

Combine gelatin, sugar and salt; mix well. Stir in boiling water and continue stirring until gelatin dissolves. Cool. Add lime rind, lime juice, carbonated beverage and sour cream; stir gently until thoroughly mixed. Chill until mixture starts to thicken; fold in mandarin oranges. Pour into 4½-5-cup mold. Chill until firm.

Ms. V. K., Hyde Co., S. D.

CHOCOLATE ICE CREAM DESSERT

18 cream-filled chocolate cookies
¼ c butter, melted
1 qt. vanilla ice cream
1 c sugar
1 (5 1/3-oz.) can evaporated milk
2 sqs. unsweetened chocolate
1 c cream, whipped and sweetened
nuts

Crush cookies. Add melted butter; blend. Pat into 9x13-inch pan, reserving some for the top. Freeze. Soften ice cream; spoon on crust. Freeze. Heat sugar, milk and chocolate in saucepan, stirring constantly. Bring slowly to boil until thickened; cool. Spoon over ice cream. Freeze. Spread whipped cream on top. Sprinkle nuts over cream. Freeze.

Miss I. O. A., Norman Co., Minn.

FROSTY STRAWBERRY SQUARES

1 c sifted flour
¼ c brown sugar
½ c chopped walnuts
½ c butter, melted
2 egg whites
2/3 c sugar
1 (10-oz.) pkg. frozen strawberries, partially thawed
2 tbsps. lemon juice
1 c whipped cream

Make a crumb crust of flour, brown sugar, nuts and butter. Spread 2/3 of the crumb mixture in 13x9x2-inch pan. Bake at 350 degrees for 20 minutes, stirring occasionally. Combine egg whites, sugar and berries. Add lemon juice and beat until mixture forms stiff peaks. Fold in whipped cream. Spoon mixture over crumbs. Top with remaining crumbs. Freeze 6 hours or overnight. Cut into squares and serve.

Mrs. M. E. P., Mountrail Co., N. D.

PUMPKIN ICE CREAM DESSERT

1 envelope unflavored gelatin
¼ c cold water
½ c pumpkin
½ tsp. salt
¼ tsp. nutmeg
1 tsp. cinnamon
¼ tsp. ginger
1 tsp. vanilla
1 qt. vanilla ice cream
Crust:
1 c ginger snap crumbs
½ c butter, melted
½ c sugar

Dissolve gelatin in water; set aside. Combine pumpkin, salt, spices and vanilla; add gelatin; cook until dissolved and smooth. Add ice cream, stirring until blended. Pat ginger crumb crust into 9-inch square pan. Fill with pumpkin mixture. Sprinkle with remaining crumb topping; freeze.

Crust: Combine crust ingredients; mix, reserving ½ cup for topping.

Mrs. I. O. A., Norman Co., Minn.

RASPBERRY SWIRL

¾ c graham cracker crumbs
3 tbsps. butter, melted
2 tbsps. sugar
3 egg yolks, beaten
1 (8-oz.) pkg. cream cheese
1 c sugar
⅛ tsp. salt
3 egg whites, stiffly beaten
1 c heavy cream, whipped
1 (10-oz.) pkg. frozen raspberries, partially thawed

Combine cracker crumbs, butter and sugar. Press lightly into well-greased 7x11x1½-inch pan. Bake in moderate oven about 8 minutes; cool thoroughly. Combine beaten egg yolks, cream cheese, sugar and salt; beat until smooth and light. Whip cream until stiff; fold into beaten egg whites. Fold cream and egg white mixture into cheese mixture. Crush raspberries to a pulp in mixer or blender. Gently swirl half fruit pulp through cheese filling; spread mixture into crust. Spoon remaining pulp over top and swirl. Freeze.

Mrs. R. M., McLean Co., N. D.

ICE CREAM COCONUT MERINGUE SQUARES

4 egg whites
1 c granulated sugar
¼ tsp. salt
½ tsp. vinegar
1 c crushed graham cracker crumbs
½ c chopped walnuts
½ c shredded coconut
vanilla ice cream

Beat egg whites; add sugar, salt and vinegar; beat until glossy. Fold in crumbs, nuts and coconut. Spread in 9-inch square pan and bake at 350 degrees for 35-40 minutes. Take from oven; cool. When cooled, spread with vanilla ice cream about ½-inch thick. Place in freezer and freeze.

Mrs. H. C. W., Polk Co., Minn.

STRAWBERRY MOUSSE

1 egg, beaten
¾ c confectioners sugar
1 tsp. vanilla
1 pt. whipped cream
1 pkg. slightly crushed, sweetened, frozen strawberries, thawed
1 tbsp. lemon juice
dash of salt

Combine egg, confectioners sugar and vanilla; beat well. Add whipped cream. Add remaining ingredients. Place in a flat cake pan with a cover and freeze.

Mrs. O. R., Spink Co., S. D.

PINEAPPLE ICE CREAM

⅛ tsp. soda
2 c sour cream
2 cans sweetened condensed milk
1 qt. whole milk
2 c crushed pineapple

Stir soda into sour cream; combine with milk. Freeze in a gallon ice cream freezer until half frozen. Add fruit and continue freezing for 30 minutes.

Mrs. K. K., Day Co., S. D.

VANILLA ICE CREAM

1 c rich milk
½ c sugar
⅛ tsp. salt
1 tbsp. flour
1 c cream
1 tbsp. vanilla

Scald milk. Gradually stir in a mixture of sugar, salt and flour. Place over low heat; boil 1 minute; cool. Pour into freezing tray and freeze ½ hour. Whip together cream and vanilla; add to frozen mixture. Freeze. Stir often during first hour.

Mrs. E. E., Cass Co., S. D.

Pudding/**DESSERTS**

HONEY STEAMED PUDDING

¼ c butter
½ c honey
1 egg, beaten
2¼ c flour
3½ tsps. baking powder
¼ tsp. salt
1 c milk
½ tsp. vanilla

Cream butter; gradually add honey. Stir in egg. Sift together dry ingredients; add alternately with milk to creamed mixture. Mix in vanilla. Pour into buttered individual molds. Cover with waxed paper held in place with a rubber band. Place molds in a steamer for 30 minutes. Serve hot.

Mrs. G. B., Beadle Co., S. D.

DESSERTS/Pudding

OLD-FASHIONED CHOCOLATE PUDDING

2½ tbsps. cornstarch
4 tbsps. cocoa
5 tbsps. sugar
dash of salt
1 tsp. vanilla
¼ c cold milk
1½ c skimmed milk

Blend together cornstarch, cocoa, sugar, salt, vanilla and cold milk. Stir into skimmed milk which has been brought to the boiling point in double boiler. Cook until thick and smooth.

Mrs. C. F. W., Adams Co., N. D.

DESSERTS/Refrigerator

ANGEL DELIGHT

1 pkg. orange gelatin
1 c hot water
½ pt. whipping cream
1 (10-oz.) can crushed pineapple
½ c nut meats
small angel food cake

Dissolve gelatin in hot water. When slightly congealed, add whipping cream and whip together. Add pineapple and nut meats. Pour mixture over angel food cake that has been broken into small pieces. Mix lightly; pour into mold. Refrigerate.

D. J., Hyde Co., S. D.

MARSHMALLOW DESSERT

1 c crushed graham crackers
2 tbsps. sugar
¼ c butter, melted
½ c milk, scalded
30 marshmallows
1 c sweet cream, whipped
1 large almond chocolate bar, cut fine
1 tsp. vanilla

Mix together cracker crumbs, sugar and butter. Press ¾ mixture into pan. Combine scalded milk and marshmallows; mix until melted. Cool. Fold in whipped cream. Add chocolate bar and vanilla; pour over crumb crust. Sprinkle remaining crumbs on top. Refrigerate.

Mrs. H. H., Jerauld Co., S. D.

FRUIT WHIP

1 c chopped well drained, cooked fruit
3-5 tbsps. confectioners sugar
dash of salt
1 tsp. lemon juice
3 egg whites, stiffly beaten

Mix and chill fruit, confectioners sugar, salt and lemon juice. Fold in egg whites; chill. May be served with sweetened whipped cream.

Mrs. H. W. Mercer Co., N. D.

APRICOT DESSERT

14 graham crackers, crushed
3 tbsps. butter, melted
½ c confectioners sugar
½ c butter
2 egg yolks
2 egg whites, beaten
1 c apricot pulp
1 c cream, whipped
2 tbsps. confectioners sugar

Mix crackers with melted butter. Press into 8x8-inch pan, reserving crumbs for top. Cream together ½ cup confectioners sugar and butter; add egg yolks. Mix well. Fold in egg whites. Pour over crumb mixture. Cover with apricot pulp, made either by cooking ½ pound dried apricots or by mashing well-drained canned apricots. Sweeten whipped cream with 2 tablespoons sugar; spread over apricots. Sprinkle with remaining crumbs. Chill 6 hours before serving.

Mrs. M. H., Emmons Co., N. D.

CHOCOLATE-COCONUT DELIGHT

1 c butter
2 c confectioners sugar
4 sqs. chocolate, melted
4 eggs
2 tsps. vanilla
1 c coconut
1 c chopped nuts
1½ c whipped cream
3 doz. vanilla wafers, crushed

Cream together butter and confectioners sugar. Add chocolate and eggs, one at a time. Beat well after each addition. Add vanilla. Stir in coconut, nuts and whipped cream. Place 2/3 crushed wafers in bottom of large pan which has been lined with waxed paper. Pour in chocolate mixture; top with remaining crumbs. Refrigerate for 24 hours.

Mrs. S. F., Ward Co., N. D.

dairy products

Milk is nature's most nearly perfect food. Children and adults need the protein, vitamins and calcium most readily available in milk and milk products. It is recommended that children have 3 or more cups of milk a day, teen-agers 4 or more and adults 2 or more.

Milk can be used as a hot or cold beverage, in gravies and on cereals. Cheese combined with spaghetti and macaroni is a nutritious main dish. And, of course, butter and everyone's favorite dessert, ice cream, help supply the needed milk requirements.

DANISH PUDDING DESSERT

1 pkg. Danish pudding, currant or raspberry
1½ c water
1 lb. fresh or frozen strawberries
1 (10-oz.) angel food loaf cake

Boil pudding and water for 1 minute; let cool 15 minutes. Add strawberries; set aside. Break up angel food cake into small pieces into a 13x9x2-inch cake pan. Cover with pudding mixture; let set until firm. Serve with whipped cream or ice cream.

Mrs. R. S., Hand Co., S. D.

PEPPERMINT MARSHMALLOW DESSERT

2½ c graham cracker crumbs
¼ c brown sugar
½ c butter, melted
1 pt. cream, whipped
1 (10½-oz.) pkg. miniature marshmallows
11 peppermint sticks, crushed
1 c chopped nuts

Combine cracker crumbs and butter; press into a 9x12-inch pan. Beat egg whites until stiff; add sugar. Spread over cracker mixture. Bake at 400 degrees for 10 minutes; cool. Fill with cherry pie mix. Refrigerate 4 to 5 hours. Serve with whipped cream or ice cream.

Mrs. W. S., Sheridan Co., N. D.

SODA CRACKER DESSERT

30 white soda crackers, crushed
¾ c butter, melted
3 egg whites, stiffly beaten
1 c sugar
1 can cherry pie mix

Combine cracker crumbs and butter; press into a 9x12-inch pan. Beat egg whites until stiff; add sugar. Spread over cracker mixture. Bake at 400 degrees for 10 minutes; cool. Fill with cherry pie mix. Refrigerate 4 to 5 hours. Serve with whipped cream or ice cream.

Mrs. V. R., Richland Co., N. D.

STRAWBERRY DESSERT

1 (10-oz.) pkg. frozen strawberries, thawed, or 1 pt. sliced, sweetened fresh strawberries
½ lb. (16) marshmallows, quartered
1 (9-oz.) can pineapple, drained
½ c coconut
½ c heavy cream, whipped

Combine strawberries and marshmallows; let stand several hours in refrigerator. Add remaining ingredients. Chill.

Mrs. O. K., Brown Co., S. D.

CRANBERRY REFRIGERATOR DESSERT

- 2 c chopped or ground fresh cranberries
- 1 large banana, diced
- 2/3 c sugar
- 2 c crushed vanilla wafers
- ½ c butter or margarine
- 1 c confectioners sugar
- 2 eggs
- ¼-½ c chopped walnuts
- 1 c heavy cream, whipped

Mix together cranberries, banana and sugar. Place 1 cup crushed wafers in bottom of 8-inch square pan. Cream together butter and confectioners sugar; add eggs, one at a time; beat well. Spread this mixture over crumbs in pan. Top with cranberry-banana mixture; sprinkle with walnuts. Spread whipped cream over cranberries and walnuts. Sprinkle with remaining cup crushed wafers. Chill overnight.

Mrs. S. J. M., Gregory Co., S. D.

JELLO CHRISTMAS CAKE

- 2 (9-inch) baked sponge cake layers
- 1 pkg. lime gelatin
- 1 pt. boiling water
- 1 pkg. cherry or strawberry gelatin
- 1 pt. boiling water
- whipped cream

Place sponge cake in layer cake pans, with foil bottoms cut to fit. Dissolve lime gelatin in boiling water; keep hot. Pour over one layer of cake, making sure that just enough liquid is used to fill cake. Repeat same method with cherry or strawberry gelatin and other cake layer. Place both in refrigerator to set overnight. Spread first layer with whipped cream; center second layer on top. Cover completely with whipped cream; serve.

Variation: Cut each layer in cubes. Toss both with whipped cream. Serve in sherbet glasses.

Mrs. F. B., Pennington Co., S. D.

PINEAPPLE RICE CUSTARD

- ½ c uncooked rice
- 1½ c boiling water
- ½ tsp. salt
- 2 c milk
- 1 c sugar
- 4 egg yolks, slightly beaten
- 4 egg whites
- 2 c diced pineapple
- ¾ c diced marshmallows
- ½ c chopped nuts
- 2 tbsps. lemon juice

Combine rice, boiling water and salt; cook for 5 minutes. Add milk and ½ cup sugar. Cook, covered, over low heat for 1 hour. Add egg yolks; cook 3 minutes longer; chill. Beat egg whites until stiff, but not dry. Gradually beat in remaining ½ cup sugar. Add remaining ingredients to rice mixture; fold in egg whites. Chill or freeze before serving. Top with whipped cream.

Ms. D. H., Ziebach Co., S. D.

DESSERTS/Refrigerator

RUMTOPF CROCK

3 medium pears, cored and chopped
3 medium oranges, peeled and chopped
1 c maraschino cherries
1 c pineapple tidbits
1 c sugar
1 c rum

In crock combine all fruits. Any fresh fruits may be added. Stir in sugar. Pour in rum. Loosely cover with cheesecloth. Refrigerate for about 2 weeks. For every cup of fruit removed, add 1 cup of fruit, ¼ cup sugar and ¼ cup rum.

Mrs. H. R., M., Mercer Co., N. D.

DESSERTS/Miscellaneous

CREAM PUFFS AND CUSTARD

½ c butter
1 c boiling water
1 c sifted flour
4 eggs

Custard Filling:
1½ c milk
¼ c sugar
1½ tbsps. cornstarch
2 egg yolks, beaten
1 tsp. vanilla

Bring to boil butter and water; add flour all at once. Stir constantly with wooden spoon until mixture clears the pan and forms a ball (about 1 minute). Remove from heat; cool. Add eggs, one at a time, beating after each addition. Beat until smooth and velvety. Drop on lightly greased cookie sheet, forming balls the size of small oranges. Bake at 450 degrees for 15 minutes; lower temperature to 350 degrees; bake 30-35 minutes longer. After cream puffs have cooled slowly, cut off tops; scoop out soft dough; fill with desired filling. Replace tops; dust with confectioners sugar.

Custard Filling: Heat milk in heavy suacepan until bubbles form around edge of pan. Combine sugar and cornstarch; stir in hot milk all at once. Cook over medium heat until mixture boils. Reduce heat; simmer 1 minute. Beat small amount of hot mixture into beaten egg yolks. Pour back into saucepan; cook, stirring until mixture boils and thickens. Stir in vanilla; refrigerate until ready to use.

Mrs. I. K., Pennington Co., S. D.

DESSERTS/Ice Cream Toppings

DUTCH HONEY

½ c brown sugar
1/3 c sugar
½ c cream
½ c corn syrup

Boil together all ingredients. Serve as a caramel topping for ice cream.

Mrs. C. W., Renville Co., N. D.

FUDGE DESSERT TOPPING

1 c chocolate chips
¼ c sugar
¼ c white corn syrup
½ c boiling water

Melt chocolate chips in double boiler; add sugar; mix well. Add corn syrup; stir until smooth. Gradually add boiling water. Cook 2 minutes. Serve either hot or cold.

Variation: For a caramel topping, use caramel chips in place of chocolate chips.

Mrs. J. E., Turner Co., S. D.

Sauces/**DESSERTS**

CHERRY SAUCE

¼ c sugar
1 tbsp. cornstarch
¾ c cherry juice
⅛ tsp. red food coloring
1 tbsp. lemon juice
1 tsp. butter
1 (20-oz.) can frozen cherries, thawed, drained

Combine sugar and cornstarch; stir in cherry juice and food coloring. Cook until mixture is thickened, stirring constantly. Remove from heat; add lemon juice, butter and cherries. Cool.

Mrs. M. L., Traill Co., N. D.

LEMON SAUCE

juice of 1 lemon
grated rind of 1 lemon
¾ c sugar
¼ c confectioners sugar
½ c butter
1 egg, beaten
dash of vanilla
dash of nutmeg

Combine all ingredients in double boiler; cook until thickened.

Mrs. F. W. H., Brown Co., S. D.

MARASCHINO SAUCE

¼ c maraschino juice
2 tbsps. lemon juice
6 cherries, chopped
2 c confectioners sugar

Mix fruit juices and cherries; add sugar and beat well.

Ms. J. K., Brown Co., S. D.

CLEAR SAUCE

½ c butter
3 tbsps. cornstarch or flour
1 c sugar
1 tsp. nutmeg
3 c boiling water

Combine all of the above ingredients. Bring to boil and boil a few minutes.

Mrs. S. H., Steele Co., N. D.

PIES 'N PASTRY

Perhaps some of the tastiest pies and pastries you'll ever find anywhere were discovered in some small Dakota farm town at the annual church bake sale. Thousands of farmwives each year enter their favorite family recipes, many a product of their own home gardens. It's "prize" recipes like these we've collected in the pages that follow. If you like pies, you will love these!

DREAMY HIGH PUMPKIN PIE

1 envelope unflavored gelatin
¼ c cold water
3 egg yolks
1 c sugar
1 1/3 c canned or fresh cooked pumpkin
1/3 c milk
1 tsp. mace
½ tsp. salt
3 egg whites
1½ c whipped cream
1 baked pie shell
toasted coconut (optional)

Soften gelatin in cold water. Beat egg yolks until thick. Add ½ cup sugar; mix well. Stir in pumpkin, milk, mace and salt. Cook over hot water until thick, stirring constantly; add to gelatin mixture. Stir until thoroughly dissolved; cool. Add ½ cup sugar to stiffly beaten egg whites; fold into pumpkin mixture. Stir in ½ cup whipped cream; spoon into baked pie shell. Chill several hours. Spread with 1 cup whipped cream; garnish with coconut.

Mrs. H. L., Mercer Co., N. D.

MILE-HIGH LEMON PIE

8 egg yolks
2 c sugar
juice of 2 lemons
grated rind of 2 lemons
2 envelopes unflavored gelatin
½ c cold water
8 egg whites
1 baked pie shell

Combine egg yolks, 1 cup sugar, lemon juice and lemon rind. Cook in double boiler until thick. Combine gelatin and cold water; add to hot pudding; cool. Beat egg whites until stiff but not dry. Add 1 cup sugar gradually to egg whites; beat again. Fold into pudding mixture. Pile into baked pie shell. Refrigerate a few hours.

Mrs. R. K., Pennington Co., S. D.

ORANGE CHIFFON PIE

1 envelope unflavored gelatin
1/3 c cold water
4 egg yolks
¾ c sugar
1 (6-oz.) can frozen orange juice concentrate
¼ tsp. salt
4 egg whites
1 baked pie shell

Soften gelatin in water. Beat egg yolks in top of double boiler until light; beat in ½ cup sugar. Add orange juice concentrate and salt. Cook over boiling water, stirring often until thickened; remove from heat. Add softened gelatin; mix well. Cool. Beat egg whites until foamy; gradually add ¼ cup sugar, beating constantly until soft peaks form. Fold into orange mixture. Pour into pastry shell; chill until set. Serve with whipped cream.

Mrs. D. T., Brown Co., S. D.

PIES/Chiffon

WHITE PIE

1 c sugar
¼ c flour
1 envelope unflavored gelatin
½ tsp. salt
1¾ c milk
¾ tsp. vanilla
¼ tsp. almond flavoring
3 egg whites
¼ tsp. cream of tartar
½ c cream, whipped
1 baked pie shell
coconut
strawberries or raspberries (optional)

Blend together ½ cup sugar, flour, gelatin and salt. Gradually stir in milk. Cook over low heat until mixture boils, stirring constantly. Boil for 1 minute. Place pan in cold water; cool until mixture mounds slightly when dropped from spoon. Blend in vanilla and almond flavoring. Beat egg whites, gradually adding cream of tartar and ½ cup sugar to make meringue. Fold first mixture carefully into meringue. Gently fold in whipped cream. Pile in baked pie shell. Sprinkle with coconut. Chill several hours until set. Top with crushed strawberries or raspberries.

Mrs. C. T., Hand Co., S. D.

PIES/Fruit

APRICOT PIE

pastry for 2-crust pie
1/3 c flour
1 c sugar
⅛ tsp. nutmeg
4 c halved, stoned apricots
2 tbsps. apricot nectar
2 tbsps. butter

Line pie plate with pastry. Refrigerate until ready to fill. Combine flour, sugar and nutmeg. Put ¼ of this mixture in bottom of pie crust. Arrange apricots, cut side up, in rings on pie crust. Add remainder of flour-sugar mixture. Add apricot nectar; dot with butter. Put lattice strips on top. Bake at 450 degrees for 10 minutes; reduce heat to 375 degrees and bake for 30 minutes longer.

Mrs. K. B., Union Co., S. D.

DATE PIE

2 eggs, beaten
¾ c brown syrup
⅛ tsp. salt
1 1/3 c sour cream
1 tsp. vanilla
1 tsp. lemon extract
1 c chopped dates
1 unbaked pie shell

Combine all ingredients. Pour into unbaked pie shell. Bake at 450 degrees for 10 minutes. Reduce heat to 350 degrees and bake for 20 minutes longer.

Mrs. M. E. P., Mountrail Co., N. D.

FROZEN DATE PIE

1 (7¼-oz.) pkg. dates, chopped
½ c water
½ c sugar
1 tsp. lemon juice
1 envelope unflavored gelatin
½ c cold water
¼ c chopped pecans

Crust:
1 (4¾-oz.) pkg. vanilla wafers, crushed
1/3 c butter or margarine, melted
1 pt. vanilla ice cream

Combine dates, ½ cup water, sugar and lemon juice. Boil until dates are tender; cool. Add gelatin softened in ½ cup cold water and pecans. Prepare pie crust by combining vanilla wafers and butter; press into pie pan. Spread ½ softened ice cream over crust. Spread date mixture over ice cream; top with another layer of ice cream. Freeze.

Mrs. J. W. D., Kingsbury Co., S.D.

PEACH CREAM PIE

¾ c sugar
2 tbsps. flour
¼ tsp. salt
1 c sour cream
1 egg, slightly beaten
½ tsp. vanilla
2 c sliced peaches, canned or fresh
1 unbaked pie shell

Topping:
1/3 c brown sugar
1/3 c flour
¼ c soft butter
½ tsp. cinnamon

Combine sugar, flour and salt. Beat in sour cream, egg and vanilla. Add peaches; pour into pastry-lined 9-inch pie pan. Bake at 400 degrees for 15 minutes; reduce heat to 350 degrees and bake 30 minutes longer. Remove from oven; spread with topping. Bake at 400 degrees for 10 minutes, or until bubbly and slightly browned.

Topping: Blend together all topping ingredients.

Mrs. J. S., Sanborn Co., S. D.

GOOSEBERRY PIE

2 c gooseberries
1 c water
1¾ c sugar
¼ c quick-cooking tapioca
1 envelope unflavored gelatin
¼ c cold water
green food coloring
1 c condensed milk or heavy cream, whipped
1 graham cracker crust
graham cracker crumbs

Cook gooseberries a few minutes in water. Add sugar; stir in tapioca and cook until clear. Dissolve gelatin in cold water; add to hot gooseberries. Put in a few drops green food coloring. Cool. Combine whipped milk or cream with gooseberry mixture. Pour into prepared graham cracker crust; top with graham cracker crumbs. Chill several hours before serving.

Mrs. S. G., Minnehaha Co., S. D.

HUCKLEBERRY PIE

1 qt. garden huckleberries
½ c brown sugar
1½ c sugar
¼ tsp. nutmeg
¼ tsp. salt
1 oz. fresh lemon or lime juice
pastry for 2-crust pie

Simmer fruit gently in small amount of water until tender; drain if necessary. Thoroughly mix remaining ingredients; pour into pie shell. Dot mixture with butter; cover with top crust. Brush top crust with cream; cut steam vents. Bake on lowest rack of oven at 500 degrees for 5 minutes. Raise pie to upper part of oven and continue baking at 350 degrees for 40 minutes, or until done.

Mrs. W. A., Grand Forks Co., N. D.

PINEAPPLE PIE

1 envelope lemon-flavored gelatin
1 c boiling water
1 pt. vanilla ice cream
1 can crushed pineapple, drained
1 baked pie shell

Dissolve gelatin in boiling water. Stir in ice cream and pineapple. Chill until partially thickened. Stir and pour into baked, cooled pie shell. Chill until firm.

Mrs. I. O. A., Norman Co., Minn.

RHUBARB PIE

1½ c sugar
3 tbsps. flour
½ tsp. nutmeg
1 tbsp. butter
2 eggs, beaten
3 c chopped rhubarb
pastry for 2-crust pie

Blend together sugar, flour, nutmeg and butter. Add beaten eggs; add rhubarb. Pour into unbaked pie shell. Cover with top crust. Bake at 450 degrees for 10 minutes; reduce heat to 325 degrees and bake for 30 minutes, or until filling thickens.

Miss. A. H., Griggs Co., N. D.

STRAWBERRY RHUBARB PIE

3 c cut-up rhubarb
1½ c sugar
1 envelope strawberry-flavored gelatin
pastry for 2-crust pie

Mix together rhubarb, sugar and gelatin. Pour into unbaked pie crust. Cover with top crust; sprinkle with sugar. Bake at 350 degrees for 45-50 minutes.

Mrs. T. McP., Grant Co., S. D.

STRAWBERRY CURRANT PIE

1 pkg. currant-raspberry Danish junket dessert
2 c cold water
2 c fresh strawberries
1 baked pie shell

Stir junket dessert into cold water. (You may use fruit juice for part of liquid.) Boil 1 minute, stirring constantly. Cool. Add strawberries; pour into baked pie shell. Chill 3-4 hours.

Mrs. F. Z., Logan Co., N. D.

VERY STRAWBERRY PIE

4 cups fresh strawberries
3 tbsps. cornstarch
¾ c sugar
2 tbsps. lemon juice
1 baked pie shell

In saucepan, crush 2 cups berries; add cornstarch and sugar. Cook stirring, until mixture boils; boil 2 minutes. Fold in lemon juice and remaining berries, halved. Pour into crust. Chill. Serve topped with whipped cream.

Mrs. A. L., Logan Co., N. D.

Meringue/**PIES**

MERINGUE

3 egg whites
3 tbsps. very hot water
5 tbsps. sugar
1 tsp. vanilla

Beat egg whites and hot water until stiff. Gradually add sugar and vanilla.

Mrs. D. H., Harding Co., S. D.

SOUR CREAM RAISIN PIE

3 egg yolks
½ c sugar
1 tbsp. flour
pinch of salt
1 tsp. cinnamon
1 c sour cream
1 c raisins
1 baked pie shell

Beat egg yoks; add sugar blended with flour, salt and cinnamon. Add sour cream and raisins. Cook over low heat until thick. Pour into baked pie shell. Cover with meringue and brown lightly in 300 degree oven.

Mrs. D. H., Harding Co., S. D.

PIES/Meringue

BUTTERMILK PIE

- 1 c sugar
- 3 tbsps. flour
- pinch of salt
- 4 egg yolks, beaten
- 2 c rich buttermilk
- 1 tsp. lemon extract
- 1 tbsp. butter
- 1 baked pie shell
- meringue

Blend together sugar, flour and salt; add to beaten egg yolks. Add buttermilk. Cook over low heat until thick. Add lemon extract and butter. Pour into baked pie shell; cover with meringue. Brown lightly.

Mrs. D.H., Harding Co., S.D.

BUTTERSCOTCH PIE

- 3 tbsps. butter
- 1 c brown sugar
- 1½ c boiling water
- 3½ tbsps. cornstarch
- ¼ c cold water
- ¼ tsp. salt
- 3 egg yolks, beaten
- 1 tsp. vanilla
- 1 baked pie shell
- meringue

Melt butter in heavy saucepan; add brown sugar. Cook over low heat, stirring constantly, until rich brwon. Add boiling water. Bring to boil again, stirring until syrup is dissolved. Slowly stir in cornstarch mixed with cold water and salt. Cook until thick; remove from heat. Gradually add egg yolks. Return to stove; cook until thick. Add vanilla; pour into baked pie shell. Top with meringue; brown in oven.

Mrs. C. G., McPherson Co., S. D.

PIES/Miscellaneous

CARAMEL PECAN PIE

- 1 1/3 c (16 oz.) sweetened condensed milk
- ½ c brown sugar
- pinch of salt
- ½ cup coarsely chopped pecans
- 1 baked pie shell
- 1 cup heavy cream
- ¼ c sifted confectioners sugar
- toasted coconut

In top of double boiler mix together milk, brown sugar and salt. Cook over rapidly boiling water, stirring often, until thickened. Remove from heat and stir in half the pecans. Turn filling into cooled pie shell. Sprinkle remaining pecans on filling. Cool to room temperature. In small mixing bowl, whip cream until stiff; fold in confectioners sugar. Pile on top of filling. Garnish with coconut. Chill.

Mrs. M. F., Hanson Co., S.D.

MOCK APPLE PIE

1½ c sugar
1½ c water
1½ tsps. cream of tartar
12 soda crackers broken up
pastry for 2-crust pie
cinnamon
butter

Boil together for 2 minutes sugar, water and cream of tartar; cool. Add soda crackers. Place in unbaked pie crust. Sprinkle with cinnamon and dot with butter. Cover with top crust. Bake at 400 degrees for 25 minutes. Reduce to 350 degrees and bake for 20 minutes longer.

Mrs. E. F. Codington Co., S. D.

CUSTARD PIE

1 1/3 cup sugar
3 tbsps. flour
½ tsp. cinnamon
¼ tsp. nutmeg
¼ tsp. cloves
4 egg yolks, slightly beaten
3 tbsps. butter, melted
1¼ c milk
4 egg whites
½ tsp. salt
1 unbaked pie shell

Mix together sugar, flour, cinnamon, nutmeg and cloves. Blend in egg yolks and melted butter. Gradually add milk; mix well. Beat egg whites with salt until stiff but not dry. Fold into the egg yolk mixture. Pour into pie shell. Bake at 350 degrees for 40-50 minutes.

Mrs. A. T., Hettinger, N.D.

POOR MAN'S PECAN PIE

1 c sugar
2 eggs
½ c margarine, melted
dash of salt
1 tsp. vanilla
1 c oatmeal
1 c dark syrup
1 unbaked pie shell

Mix together all ingredients; pour into unbaked pie shell. Bake at 350 degrees for 45 minutes.

Miss D. B., Stutsman Co., N. D.

CARROT PIE

½ c sugar
1 c mashed carrots
1 c milk
1 tsp. cornstarch
2 eggs, beaten
1 tsp. cinnamon
½ tsp. cloves
¼ tsp. ginger
½ tsp. salt
single pie crust

Combine all ingredients; mix well. Bake at 450 degrees for 15 minutes. Reduce heat to 375 degrees; bake for 45 minutes longer.

Mrs. O. K., Brown Co., S. D.

BAKED TYPE TURNOVERS

1 c flour
dash of salt
1 egg yolk, beaten
¼ c milk
3 tbsps. lard

Sift together flour and salt. Combine egg yolk and milk. When well-blended, add flour and salt. Mix well; roll out ⅛-inch thick. Spread 1 tablespoon lard over this. Fold in half; refrigerate 15 minutes. Repeat folding and chilling two more times, using 1 tablespoon lard each time. Roll out; cut in squares; fill with your favorite filling. Fold squares in triangles; brush tops with water; sprinkle with sugar. Bake at 400 degrees for 20 minutes.

Mrs. E. V. S., Miner Co., S. D.

NEVER-FAIL PIE CRUST

1¼ c lard
3 c flour
1 tsp. salt
1 egg, well-beaten
5 tbsps. water
1 tbsp. vinegar

Cut lard into flour and salt. Combine egg, water and vinegar. Pour liquid into flour mixture; blend with spoon until flour is moistened. This crust can be re-rolled without toughening and keeps in the refrigerator up to 2 weeks. Makes enough crust for 2 double-crust pies.

Miss M. S., Beadle Co., S. D.

MEATS & POULTRY

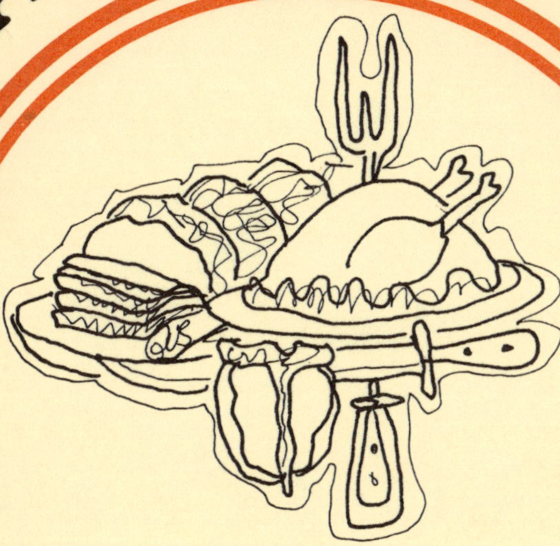

Big appetites are a by-product of hard work. Therefore, farmwives regularly prepare three hearty meals a day for their busy families. Meat is the mainstay of these meals. This section features a large selection of rib-sticking meat and poultry dishes, from taste-tempting steaks to economical chicken casserole.

Beef
Ground Beef
Lamb
Pork
Miscellaneous
Sauces
Chicken
Turkey
Poultry Stuffing

CALIFORNIA CASSEROLE

2 lbs. steak
1/3 c flour
1 tsp. paprika
¼ c cooking oil
½ tsp. salt
⅛ tsp. pepper
1 c water
1 can cream of chicken soup
1¾ cans water
1¾ c (No. 2 can) tiny onions

Crumb Dumplings:
2 c flour
4 tsps. baking powder
½ tsp. salt
1 tsp. poultry seasoning
1 tsp. celery seed
1 tsp. onion flakes
1 tbsp. poppy seeds
¼ c salad oil
1 c milk
¼ c butter, melted
1 c bread crumbs

Coat steak with flour and paprika. Brown in mixture of oil, salt, pepper and water (part onion liquid may be used). Cover and simmer for 30 minutes or until tender. Transfer meat to a 3-quart casserole or 14x10-inch pan. Heat chicken soup in skillet used to brown meat. Blend in water; bring to a boil stirring constantly. Pour over meat. Top with crumb dumplings. Bake at 425 degrees for 20-25 minutes.

Crumb Dumplings: Sift together flour, baking powder, salt, poultry seasoning, celery seed, onion flakes and poppy seed. Add salad oil and milk; stir until just moistened. Drop rounded tablespoon of dough into butter then into bread crumbs, coating well.

Mrs. A. J., Hand Co., Mont.

CORNISH PASTRIES

Crust:
2 c lard
6 c flour
1 tbsp. salt
water

Filling:
3 lbs. beefsteak, cubed
1 c suet, ground
2 c diced onions
3 lbs. potatoes, cooked and diced
¾ tsp. black pepper
1 tbsp. salt
1½ tsps. monosodium glutamate

Combine crust ingredients; divide into 3 batches. Roll out and cut in 6-inch diameter circles. Chill while preparing filling.

Filling: Combine beef, suet, onions and potatoes. (Drain potatoes and pat dry before adding.) Add seasonings; mix thoroughly. Spread filling evenly on crusts, keeping away from edges. Moisten lower edge of crust with water; fold top half over bottom; seal edges together by pressing with fork. Place on cookie sheet; bake at 400 degrees for 20 minutes; reduce heat to 325 degrees; bake 45 minutes longer.

Mrs. F. B., Mercer Co., S. D.

SWISS STEAK

- 1 large slab round steak, ½-in. thick
- flour
- salt
- pepper
- monosodium glutamate
- 1 large onion, sliced thin
- 1 qt. tomatoes
- 1 can cream of mushroom soup

Cut steak in serving pieces. Dredge on both sides in flour and seasonings. Pound well. Brown meat in small amount of fat in hot skillet. Transfer to roaster and alternate layers of meat with onion. Combine tomatoes and soup; pour over meat. Bake at 350 degrees for 2 hours.

Mrs. J. J. L., Sargent Co., N. D.

STEAK 'N POTATO SUPPER

- 1½ lbs. round steak
- flour
- salt
- pepper
- 4 medium potatoes, pared and thinly sliced
- 2 medium onions, thinly sliced
- 2 beef bouillon cubes dissolved in 1½ c boiling water or 1 can condensed beef broth

Cut steak in serving size pieces. Coat in flour, salt and pepper mixture; slowly brown in hot fat (25-30 minutes). Cover tightly and simmer until tender, adding water if necessary. Layer potatoes and onions over meat; season. Dissolve bouillon in boiling water; pour over meat. Cover tightly and cook slowly until potatoes are tender (about 35 minutes). Add water if necessary.

Mrs. D. N., Stark Co., S. D.

QUICK BEEF STEW

- 2 lbs. beef chuck, cubed
- 2 tbsps. shortening
- ¼ c chopped onion
- ½ c minced green pepper
- ¼ tsp. whole allspice
- 1 tsp. salt
- 2 cans tomato soup
- 6 baked potatoes

Brown meat in shortening in a heavy skillet. Add onions, green pepper, allspice and salt. Pour in soup; cover and simmer about 45 minutes or until done. Serve over hot baked potatoes.

Mrs. G. J., Butte Co., S. D.

BEEF JERKY

Cut 3 pounds top round USDA Choice into very thin strips. Remove all fat from meat. Put ½-inch thick layer of beef in stone crock; sprinkle with salt, pepper and liquid barbecue smoke. Repeat, using all the beef. Weight down; leave overnight. Drain; dry with paper towel. Place strips on racks in oven. Bake at 150 degrees for 4-11 hours. (Length of time you heat meat depends on how dry you like your jerky.)

Mrs. M. B., Brown Co., S. D.

CHICKEN FRIED ROUND STEAK

1½ lbs. round steak, ½-in. thick
1 egg, beaten
1 tbsp. milk
2/3 c fine cracker crumbs
¼ c salad oil
salt
pepper

Pound steak thoroughly; cut in serving pieces. Blend egg and milk. Dip meat in egg mixture, then in crumbs. Slowly brown meat in hot oil, turning once. Season with salt and pepper. Cover; cook over low heat for 30-45 minutes or until tender.

Mrs. R. T., Norman Co., Minn.

Ground Beef/MEATS

BEEF AND EGGPLANT CASSEROLE

2 tsps. butter, melted
½ c bread crumbs
3 tbsps. butter
¼ c minced onion
1 lb. ground beef
2 medium-sized eggplants, pared and cubed
dash of salt, pepper and thyme
1 can tomato soup
1 c evaporated milk

Toss together 2 teaspoons melted butter and bread crumbs; set aside on waxed paper. Brown onion and beef in 3 tablespoons butter. Add eggplant and seasonings. Cook over low heat for 10 minutes. Stir in soup and milk. Pour mixture into casserole; top with buttered crumbs. Bake until crumbs are brown.

Mrs. R. A. T., Potter Co., S. D.

HAMBURGER-CORN CASSEROLE

2 lbs. ground beef
1 c chopped onion
2 c whole kernel corn
1 can cream of mushroom soup
1 can cream of chicken soup
1 c dairy sour cream
1 tsp. monosodium glutamate (optional)
1 tsp. salt
¼ tsp. pepper
¼ c chopped pimiento
3-4 c cooked noodles
1 c bread crumbs
3 tbsps. butter, melted

Brown meat; add remaining ingredients, except crumbs and butter. Place in 5-quart casserole. Top with mixture of bread crumbs and butter. Bake at 350 degrees for 30 minutes.

Note: Be careful not to overbake, or sour cream will make casserole bitter.

Mrs. W. B., LaCrosse Co., Wis.

CHOW MEIN HOT DISH

1 lb. ground beef
chopped onion
1 can cream of mushroom soup
1 can chicken vegetable soup
½ can milk
1 large can chow mein noodles

Brown ground beef and onion. Place in casserole; add remaining ingredients. Bake at 350 degrees for 20-25 minutes.

Mrs. E. S., McCook Co., S. D.

BEEF BALLS

½ c uncooked rice
1 lb. ground beef
1½ tsps. salt
¼ tsp. pepper
2 tbsps. minced onion
2 tbsps. fat
1 can condensed tomato soup
1 c hot water

Mix together rice, ground beef, salt, pepper and onion. Shape into 12 balls. Brown balls in fat in heavy fry pan. Add soup and water. Cover and cook 1½ hours or until rice is tender.

Mrs. E. S., McCook Co., S. D.

BEEF TACOS

½ c flour
½ tsp. salt
1 lb. hamburger
3 c tomato puree
½ tsp. marjoram
3 tbsps. oil
12 corn tortillas
2 c oil or lard
1 head lettuce, shredded
½ c shredded radishes
½ c peas, cooked or canned
½ lb. grated cheese

Work flour and salt into hamburger. Make sauce from tomato puree, majoram and 3 tbsps. oil; mix thoroughly. Warm tortillas if they are not soft enough to handle. Place 1½ tablespoons prepared hamburger on each tortilla; fold over; press tightly together; pin with toothpick. Fry in deep fat until crisp; drain on absorbent paper. Remove toothpicks; open tacos while still hot. Fill with shredded lettuce, sauce mixture, radishes and cooked peas. Sprinkle generously with grated cheese.

Mrs. W. J. N., Pennington Co., S. D.

WHEAT FLOUR TORTILLAS

2 c flour
1½ tsps. salt
2 tbsps. shortening
2/3 c cold water

Sift together flour and salt. Add shortening; work well into flour mixture. Add cold water; form into ball until bowl is cleaned of all dough. Knead mixture. Make dough into balls the size of an egg. Rub lard on each ball; allow to stand 10 minutes. Pat or roll out until tortillas are the size of dinner plate. Cook in ungreased skillet or frying pan; turn until brown on both sides.

Mrs. W. J. N., Pennington Co., S. D.

EASY LASAGNE

1 lb. ground beef
2 cloves garlic, minced
1 (6-oz.) can tomato paste
1 (1-lb. 4-oz.) can tomatoes
1 tsp. salt
¾ tsp. pepper
½ tsp. oregano
1 (8-oz.) pkg. lasagne noodles
1½ c cut-up Swiss cheese
1 (12-oz.) carton cottage cheese

Brown ground beef and garlic in small amount of fat. Add tomato paste, tomatoes, salt, pepper and oregano. Cover; simmer for 20 minutes. Cook noodles as directed on package. Heat oven to 350 degrees. Alternate layers of meat sauce, noodles and cheese in 11½x7½x1½-inch baking dish, beginning and ending with meat sauce. Bake for 20-30 minutes. Sprinkle with grated Parmesan cheese.

Ms. M. N., Ottertail Co., Minn.

SPAGHETTI 'N MEAT BALLS

Sauce:
7 c tomatoes
¼ c instant minced onion
¼ c salad oil
½ tsp. ground black pepper
½ tsp. basil leaves
½ tsp. garlic powder
1 tsp. sugar
2 tbsps. salt
1 bay leaf
1 (6-oz.) can tomato paste

Meat Balls:
1 lb. ground lean beef
1 c soft bread crumbs
1 tbsp. dried parsley flakes
1 tbsp. grated Parmesan cheese
¼ tsp. ground black pepper
⅛ tsp. garlic powder
1 tsp. salt
1 egg, beaten
2 tbsps. olive oil

Combine sauce ingredients; simmer all day. Combine meat ball ingredients, except olive oil; shape into 1-inch balls. Brown on all sides in hot olive oil. Add to spaghetti sauce. Serve hot over spaghetti noodles.

Mrs. R. W., Richland Co., N. D.

POTATO TOPPED BEEF LOAF

1 lb. ground beef
1 c cottage cheese
1½ c soft bread crumbs
½ c tomato juice
1 egg
½ onion, finely chopped
¼ c finely chopped celery
1 tsp. soy sauce
1½ tsps. salt
pepper
hot mashed potatoes

Combine all ingredients except potatoes in bowl; blend. Pack into buttered 8-inch square baking pan. Bake at 350 degrees for 1 hour. Top with potatoes, brush with melted butter and place under hot broiler until lightly browned.

Mrs. H. C., Meade Co., S. D.

MEXICAN TAMALE CASSEROLE

1½ lbs. ground beef
½ c chopped onion
½ c diced celery
2 tbsps. lard or drippings
1 (16-oz.) can stewed tomatoes
1 (9-oz.) can whole kernel corn, drained
1 tsp. salt
1 tsp. oregano
1 tsp. chili powder
1/3 c yellow corn meal
1 small can ripe olives, sliced and drained
grated Parmesan cheese

Lightly brown ground beef, onion and celery in lard or drippings. Add tomatoes (including liquid), corn, seasonings and corn meal. Cook clowly, stirring occasionally, for 10 minutes. Add olives. Place mixture in greased casserole; sprinkle with cheese. Bake at 350 degrees for 25-30 minutes.

Mrs. J. S., Ward Co., N. D.

JUICY MEAT LOAF

1½ lbs. ground beef
2 eggs
¾ c quick oats
1 small onion, chopped
2 tsps. salt
1 c tomato juice
½ tsp. sage

Topping:
¼ c ketchup
2 tsps. vinegar
1 tsp. Worcestershire sauce
¼ tsp. chili powder
¼ c warm water

Combine meat loaf ingredients; pat into loaf pan. Mix together topping ingredients; pour over meat loaf. Bake at 350 degrees for 1 hour. Cover pan with foil until last 15 minutes of baking.

Mrs. M. S., Brown Co., S. D.

MEAT LOAF WITH DRESSING

2 lbs. ground beef
½ c milk
1 egg, well-beaten
2 tbsps. chopped onion
⅛ tsp. pepper
1½ tsps. salt
½ c ketchup

Dressing:
3 c soft bread crumbs
2 tbsps. minced onion
1 tsp. salt
⅛ tsp. pepper
¼ tsp. sage
½ c milk

Combine meatball ingredients; mix well. Combine dressing ingredients. Spread meat mixture in a rectangle on a board. Cover with dressing. Roll up jelly roll fashion and bake until done.

Mrs. G. B., Adams Co., N. D.

MEAT BALLS IN SOUR CREAM

1½ lbs. ground beef
½ c commercial sour cream
2 tsps. salt
¼ tsp. pepper
½ tsp. garlic powder

Gravy:
1 c commercial sour cream
¼ tsp. garlic powder
½ tsp. sugar
2 tsps. dried dill

Mix meat ball ingredients thoroughly. Shape into balls; chill 15 minutes. Brown meat balls in butter; pour off excess fat. Place in baking dish; bake at 275 degrees for 10 minutes. Top with gravy.

Gravy: Add sour cream to skillet in which meat was browned. Season with garlic, sugar and dill. Heat until bubbly.

Mrs. D. G., Rolette Co., N. D.

PIGS IN A BLANKET

3-4 qts. cooked rice
1 medium onion, diced
1 lb. ground beef, crumbled
1 head of cabbage
butter
1 can tomato paste (may also use juice, soup, fresh or canned tomatoes)

Brown hamburger and onion; add to rice. Wilt the head of cabbage. Using one leaf, put desired amount of rice mixture in it, along with a dab of butter. Roll up, sealing mixture inside. Place last-lap down in a pan or kettle. Cover all with tomatoes; simmer until cabbage is tender, about ½ to 1 hour. Serve with salt and pepper or vinegar.

Note: Corned beef can be substituted for hamburger.

Ms. H. B. Walsh Co., N. D.

beef

The beef cattle industry has proved that it plays a big role in Dakota farm economy by accounting for a large part of the total income received from the sale of farm products.

Beef has always been a mainstay in the diets of Dakotans. The many beef organizations in the Dakotas, such as the North Dakota Stockmen's Association, the South Dakota Stock Growers Association and their companion Cowbells Association, suggest you eat beef often, whether it's for breakfast, lunch or dinner.

MEATS/Ground Beef

TATER TOT CASSEROLE

1½-2 lbs. ground beef
salt and pepper
sliced onion
2 pkgs. tater tots
2 cans cream of celery, mushroom or tomato soup
water

Pat ground beef into 9x13-inch pan; season with salt and pepper. Place sliced onions on ground beef. Layer tater tots over this. Mix 2 cans soup with 1 can water and pour over casserole. Bake uncovered at 350 degrees for 1 hour.

Note: Raw potatoes may be susbtituted for tater tots. Bake approximately 2½ hours.

Ms. L. K., Barnes Co., N. D.

FRENCH SUPPER LOAF

2/3 c evaporated milk
1½ lbs. ground beef
½ c cracker crumbs
1 egg
1 tbsp. mustard
½ c chopped onions
1½ tsps. salt
⅛ tsp. pepper
2 c grated cheese
1 loaf French bread

Mix together all ingredients except French bread. Cut French bread in half lengthwise. Place meat mixture on each half; wrap in foil, leaving meat mixture uncovered. Bake at 350 degrees for 25 minutes.

Mrs. D. L., Buffalo Co., S. D.

MEATS/Lamb

COUNTRY LAMB LOAF

3 lbs. lean ground lamb
1 onion, chopped
1 green pepper, chopped
1¼ c bread crumbs
1 (8-oz.) can tomato sauce
¼ c ketchup
2 eggs
2 tsps. salt
1½ tsps. oregano
1½ tsps. steak sauce
1 tsp. Worcestershire sauce
½ tsp. pepper
½ c ketchup
tomato slices
onion slices
green pepper slices

Combine all ingredients except ½ cup ketchup, tomato, onion and green pepper slices; mix well. Pack into 9x5x3-inch loaf pan. Bake at 350 degrees for 1½ hours, or until done. Remove from oven and turn onto serving platter. Heat remaining ketchup; spoon over top of loaf. Decorate with tomato, onion and green pepper slices.

Mrs. Y. H., Kingsbury Co., S. D.

LAMB CURRY BAKE

- 2 c diced cooked lamb
- 1 tsp. onion juice
- ½ c sliced celery
- ½ tsp. salt
- ¼ tsp. pepper
- 1/3 c milk
- 1 (10½-oz.) can cream of mushroom soup
- 1½ c flour
- ½ tsp. salt
- 2 tsps. curry powder
- ½ c shortening
- 3 tbsps. water

Combine lamb, onion juice, celery, ½ teaspoon salt, pepper, milk and soup in lightly greased 10-inch pie plate. Sift together flour, ½ teaspoon salt and curry powder; cut in shortening. Add water; stir until well mixed. Roll out on lightly floured surface to ⅛-inch thickness. Place over lamb mixture. Seal and flute edge; slit top. Bake at 400 degrees for 35 minutes, or until lightly browned.

Mrs. D. L., Kingsbury Co., S. D.

LAMB SHISH KEBABS

- 2 lbs. boneless lamb cubes
- 3 tbsps. oil
- ¼ c lemon juice
- 1½ tsp. salt
- 1 tsp. marjoram
- ½ tsp. oregano
- 1/3 c chopped onion
- 1 garlic clove, crushed
- 3 green peppers, cut into eighths
- 24 cherry tomatoes
- 24 mushroom caps

Trim fat from lamb. Marinate overnight in refrigerator in mixture of oil, lemon juice, salt, marjoram, oregano, onion and garlic. Pour enough boiling water over green pepper pieces to cover; let stand 5 minutes; drain. Alternate lamb cubes, green pepper pieces, whole tomatoes and mushroom caps on metal skewers. Brush with remaining marinade. Broil, turning as needed, until done (15-20 minutes).

Mrs. Y. H., Kingsbury Co., S. D.

HONEY-MUSTARD LEG ROAST

- 1 5-lb. leg of lamb
- ½ c prepared mustard
- ½ c honey
- 1 tsp. salt
- ¼ tsp. pepper

Place lamb on rack in shallow roasting pan. Bake at 300 degrees for 2 hours. Combine mustard, honey, salt and pepper; blend. Pour over lamb. Bake for 1-2 hours, or until meat thermometer registers 175-180 degrees.

Mrs. J. C., Morton Co., N. D.

LAMB SHOULDER CHOPS

4 lamb shoulder chops
1 tbsp. salad oil
1 tbsp. flour
1 (1½-oz.) pkg. spaghetti sauce mix
2 c water
1 (8-oz.) can tomato sauce
½ tsp. oregano
hot cooked spaghetti

Brown lamb chops in oil on both sides. Drain off all but 1 tablespoon drippings; blend in flour. Gradually add spaghetti sauce mix, combined with water, tomato sauce and oregano in skillet, stirring constantly. Heat to boiling. Reduce heat; simmer, covered, for 20 minutes. Serve chops with sauce over hot cooked spaghetti.

Mrs. Y. H., Kingsbury Co., S. D.

BAKED LAMB STEW

1½ lbs. lamb shoulder, cubed
1 c sliced onion
1½ c sliced beets
1½ c cut green beans
2 c diced tomatoes
1 c sliced mushrooms
3 c stock or bouillon
salt and pepper
1½ c biscuit mix
½ c yellow corn meal
½ c milk

Combine lamb and onions. Cook over low heat until lamb is browned on all sides. Add beets, green beans, tomatoes, mushrooms, stock or bouillon and salt and pepper; mix well. Turn into 3-quart casserole. Cover; bake at 350 degrees for 1 hour, or until lamb and beets are tender. Combine biscuit mix and corn meal; add milk; mix lightly. Turn out on lightly floured surface; knead gently 10 times. Roll out to ½-inch thickness. Cut into 2½-inch rounds, using floured cutter. Arrange biscuits over stew. Bake at 400 degrees for about 15 minutes, or until biscuits are done.

Mrs. K. M., Aurora Co., S. D.

sheep

Sheep producing is an important enterprise in many areas of the Dakotas — South Dakota ranks third in the nation in total production. However, lamb does not enjoy the popularity at the dinner table that beef and pork do. One reason may be that many housewives are unsure how to prepare and serve this meat. We encourage you to try our lamb recipes and let lamb lend delicious variety to your meals.

SAVORY LAMB STEW

- 2 tbsps. butter or margarine, melted
- 1½ lbs. lamb shoulder, cubed
- 1 medium onion, sliced
- ½ c diced celery
- 1 small clove garlic, finely chopped
- 1 medium green pepper, diced
- 2 c stock or bouillon
- 1½ tsps. salt
- ¼ tsp. pepper
- 2½-3 c diced potatoes
- 1½ c sliced carrots
- ¼ c flour
- ½ c water

Saute lamb in melted butter or margarine over low heat, stirring occasionally, until browned on all sides. Add onion, celery, garlic and green pepper; cook 5 minutes. Add stock or bouillon, salt and pepper. Cover and cook 15 minutes. Add potatoes and carrots; cook 45 minutes longer or until vegetables are tender. Remove lamb and vegetables. Combine flour and water; blend. Add flour mixture to stock or bouillon mixture; blend. Cook over low heat, stirring constantly, until thickened. Add vegetables and lamb.

Mrs. Y. H.., Kingsbury Co., S. D.

Pork/**MEATS**

HAM LOAF

- 2 c chopped, cooked ham
- 2/3 c boiled rice
- 2 egg yolks
- 1 c milk
- 2 tbsps. ketchup
- 2 tbsps. chopped onion
- ¼ tsp. salt

Creamed Egg Sauce:

- 3 tbsps. butter
- 4 tbsps. flour
- 2 c milk
- 1 hard boiled egg, diced
- 2 tbsps. chopped sweet pickle
- 1 tsp. finely chopped parsley
- ¼ tsp. paprika
- ¼ tsp. celery salt
- ¼ tsp. salt

Combine ham loaf ingredients; mix well. Pour into buttered loaf pan and bake at 350 degrees for 30-45 minutes. Unmold and surround with creamed egg sauce.

Creamed Egg Sauce: Melt butter; add flour and mix until all lumps are gone. Add milk slowly and cook until it forms a creamy white sauce. Add remaining ingredients and cook 1 minute. Pour mixture on ham loaf slices.

Mrs. L. S., Cavalier Co., N. D.

ASPARAGUS HAM HOT DISH

4 c soft bread cubes
1 c grated cheese
¼ c butter, melted
1 (No. 2) can aspargus, drained
1 tbsp. butter
6 tbsps. flour
2 c milk
¼ c chopped onion
salt
pepper
2 c diced ham

Mix together bread cubes, cheese and melted butter. Place half of mixture in bottom of 2-quart casserole. Pour asparagus over top. Make a white sauce from butter, flour, milk, onion, salt and pepper. Add ham to sauce and pour over asparagus. Top with remaining bread cubes. Bake at 350 degrees for 45 minutes.

Mrs. W. A., Barnes Co., N. D.

HAM AND VEGETABLE HOT DISH

3 c thinly sliced potatoes
1 c thinly sliced carrots
1 onion, thinly sliced
3 c diced ham
2 tbsps. flour
1 c cheese soup
1 c milk
dash of pepper

Place in 2-quart casserole potatoes, carrots, onion and ham. Blend together flour, soup, milk and pepper. Cook, stirring until mixture boils. Pour mixture over vegetables in casserole. Bake covered at 350 degrees for 1 hour. Remove cover and bake an additional 15 minutes, or until vegetables are tender.

Mrs. J. G., Walsh Co., N. D.

NOODLED HAM

2 c dry noodles
2 c cooked ham, chopped
2 eggs, beaten
1 c cream
bread crumbs
butter

Cook noodles according to package directions. Alternate layers of noodles and ham in buttered 2-quart casserole. Combine eggs and cream; pour over noodles and ham. Top with a thin layer of bread crumbs and dot with butter. Bake at 350 degrees for 35-40 minutes.

Mrs. M. E. P., Mountrail Co., N. D.

CRAB APPLE GLAZE

1 (16-oz.) jar crab apples
1/3 c brown sugar
1 tbsp. lemon juice

Drain crab apples; combine liquid with brown sugar and lemon juice in saucepan. Bring to boil, stirring to dissolve sugar. Cook slowly for 5-8 minutes. Garnish ham with heated crab apples.

Mrs. G. B., Jones Co., S. D.

SPICY MUSTARD GLAZE

½ c sugar
1/3 c apricot nectar or other fruit juice
1 tbsp. dry mustard
¼ tsp. black pepper
½ tsp. allspice
⅛ tsp. ground cloves

Combine all ingredients; stir until well blended. Spoon over hot ham.

Mrs. C. S., Dunn Co., N. D.

PINEAPPLE GLAZE

2 tbsps. brown sugar
1 tbsp. cornstarch
1 (13-oz.) can crushed pineapple
¼ tsp. ginger
¼ tsp. garlic salt
¼ c soy sauce

Combine brown sugar and cornstarch. Stir in crushed pineapple, ginger, garlic salt and soy sauce; cook 5 minutes, stirring occasionally.

Mrs. W. D., Cavalier Co., N. D.

SPICY PLUM SAUCE

1 (1-lb., 14-oz.) can purple plums
3 tbsps. cornstarch
½ tsp. salt
¼ tsp. cinnamon
⅛ tsp. allspice
2 tbsps. lemon juice
1 tbsp. butter or margarine

Drain plums, reserving syrup. Cut plums in half; remove pits. Combine cornstarch, salt, cinnamon and allspice. Add to plum syrup; stir well. Cook, stirring constantly, until thickened and clear. Add plum halves, lemon juice and butter or margarine. Cook until heated through.

Mrs. I. T., Codington Co., S. D.

APPLE-RAISIN STUFFED PORK CHOPS

6 pork rib chops
¼ c milk
1 c soft bread crumbs
¼ tsp. salt
¼ tsp. sage
1 c finely chopped pared apples
½ c raisins
1 tbsp. butter or margarine, melted
2 tbsps. lard or drippings
1 tsp. salt
⅛ tsp. pepper

Make a pocket in each chop by cutting along the bone. Pour milk over bread crumbs. Stir in ⅛ tsp. salt, sage, apple, raisins and butter or margarine. Stuff each chop with ¼ cup apple mixture. Brown chops on both sides in lard or drippings. Pour off drippings. Season chops with 1 teaspoon salt and pepper. Cover tightly; cook slowly 45 minutes to 1 hour or until done.

Mrs. M. F., Grand Forks Co., N. D.

JIMMY STEWART'S FAVORITE PORK CHOPS

- 6 pork chops
- 6 tbsps. ketchup
- 6 tbsps. brown sugar
- 6 slices lemon

In a casserole dish, layer pork chops. Top each with 1 tablespoon ketchup, 1 tablespoon brown sugar and 1 slice lemon. Bake at 350 degrees for 1 hour.

Mrs. J. L., Brown Co., S. D.

BARBECUED SPARERIBS

- 1 large onion, diced
- 1 tbsp. butter, melted
- 1 c canned tomatoes
- 1 c diced celery
- 1 c diced green pepper
- 1 c ketchup
- 2 tbsps. brown sugar
- 3 dashes of Tabasco sauce
- ½ tsp. dry mustard
- 2 c beef stock or 2 bouillon cubes dissolved in 2 c boiling water
- salt
- pepper
- 1 clove garlic
- 4 lbs. spareribs

Brown onion in butter. Add remaining ingredients, except garlic and spareribs; stir and cover. Simmer over low heat for 1 hour. While sauce is simmering, rub spareribs with cut side of garlic clove. Place in shallow baking pan; roast uncovered at 350 degrees for 30 minutes. Pour sauce over ribs; continue baking for 45 minutes longer, basting frequently.

Mrs. K. B., Cavalier Co., N. D.

pork

Pork ranks second only to beef in the amount of meat consumed in the United States. The majority of cured pork, notably bacon and ham, is eaten as a breakfast meat. With farm folk, several strips of crisp, lean bacon or a thick slice of ham provides plenty of energy to carry them through a morning of hard work. Pork chops are a typical American dish and roast pork is a delicacy the world over.

CORNED BEEF HOT DISH

- 1 can corned beef
- 1 can milk
- 1 can cream of chicken soup
- ¼ c diced cheese
- 2 c cooked macaroni
- 1 c corn (optional)

Combine all ingredients in casserole dish. Bake at 350 degrees for 45-60 minutes.

Mrs. E. E., Cass Co., N. D.

POTATO PIGGIES

- 2 c cooked mashed potatoes
- 8 frankfurters, halved crosswise
- ½ c corn flake crumbs
- ½ c dark corn syrup
- 1 c canned French fried onion rings
- parsley flakes (optional)

Spread potatoes around each frankfurter half with wide spatula or knife. Roll in corn flake crumbs. Arrange in lightly greased 8-inch square baking dish. Drizzle with syrup. Sprinkle with onion rings and parsley flakes. Bake, uncovered, at 350 degrees for 20-25 minutes, or until well browned and crisp.

Miss P. Z., Dewey Co., S. D.

FRANK-BACON BATS

- 16 frankfurters
- 1½ c pickle relish or chopped sweet pickles
- 1 tbsp. prepared mustard
- 16 slices bacon

Make lengthwise slits in frankfurters. Mix pickle relish and mustard. Spoon into slits in frankfurters. Wrap each with bacon slice and fasten with toothpicks. Place on baking sheet. Bake at 425 degrees, turning occasionally, for about 30 minutes, or until frankfurters are browned and bacon is cooked.

Miss P. K., Campbell Co., S. D.

FRANK FRITTERS

- 1 pkg. frankfurters
- 1 c pancake mix
- 1 tbsp. sugar
- 2/3 c water

Mix pancake mix, sugar and water. Dip franks in batter. Fry in 1-inch deep fat at 375 degrees for 2-3 minutes, or until brown. Drain. Insert skewers.

Mrs. S. D., Billings Co., N. D.

SUMMER SAUSAGE

8 lbs. pork
12 lbs. cured beef
3 ozs. white pepper
¼ lb. salt
½ oz. whole pepper
½ oz. mustard seed
1 oz. coriander seed

Put meat through grinder; add spices. Mix until evenly seasoned. Spread out in a cool place and leave for 36 hours. Stuff mixture into casings and let hang 24 hours. Smoke in electric smoker 12 hours at low temperature (not over 100 degrees). Hang up until thoroughly dry (about 2 weeks). It is then ready to use.

Casings for summer sausage: To make casings use strips of muslin 8-in. wide and 19-in. long (this allows for ½-in. seam). Fold and sew up one end and the side. Turn inside out. Stuff tightly with sausage, making sure there are no air holes which will cause mold on the inside of the casing. Tie the top with heavy cord and allow for a loop to hang up the sausage.

Note: This can be made in cold weather only, but will keep indefinitely (after curing) in a cool, dry place. If casing becomes moldy just wipe it off before using.

Mrs. P. K., Polk Co., Minn.

BLOOD SAUSAGE

1 hog's head
onion
salt
pepper
1 loaf moistened bread
2 lbs. raisins
2 tbsps. salt
2½ tbsps. nutmeg
1½ tbsps. cinnamon
4 tbsps. vanilla
1 c or more sugar
3-4 c blood

Cook hog's head with onion, salt and pepper until it separates from bones. Grind with moistened bread and raisins; add remaining ingredients. Stuff in regular casings; tie ends. Cook about 10 minutes.

Note: Use additional sugar if hog's head is extra large. Do not stuff casings too tightly, as meat expands about ¼ when cooked.

Mrs. A. K., Davison Co., S. D.

BOLOGNA

50 lbs. beef
1 lb. salt
1-3 lb. smoked salt
20 lbs. pork
½ c pepper
2 heads garlic
4 lbs. potato flour
2 gals. water

Cut beef in small pieces; add salts; mix well. Let stand 6 days. Grind together beef and pork; add remaining ingredients. Fill casings and smoke.

Mrs. J. S., Bon Homme Co., S. D.

LIVER SAUSAGE

1 hog's head
2 small onions
salt
pepper
ground allspice
1½ lbs. liver

Boil hog's head, trimming out most of fat. Grind meat and onions until fine; add seasoning to taste. Boil liver until there is no trace of blood when liver is pierced with a fork. Grind until fine; add to ground head meat. Stuff in casings; tie into rings. Boil slowly 10 minutes.

Note: If canning in jars, omit onions; fill jars to neck and process 3½ hours in hot water bath.

Mrs. E. S., Hamlin Co., S. D.

Sauces/**MEATS**

BLENDER BARBECUE SAUCE

15 ripe tomatoes
  (about 5 lbs.)
1 large onion
1 large green pepper
4 cloves garlic
2 c water
½ c vinegar
½ c Worcestershire sauce
½ c brown sugar
4 tsps. salt
4 tsps. chili powder
4 tsps. dry mustard
2 tsps. celery salt
½ tsp. Tabasco sauce

Liquefy tomatoes in blender. Grind up onion, green pepper and garlic in blender. Combine all ingredients; bring to boil; simmer and cook until thick, stirring occasionally. Pour into sterile jars.

Mrs. H. B., Jones Co., S. D.

TEXAS BARBECUE SAUCE

2 tbsps. brown sugar
1 tbsp. paprika
1 tsp. salt
1 tsp. dry mustard
¼ tsp. chili powder
1/3 tsp. cayenne pepper
2 tbsps. Worcestershire sauce
¼ c vinegar
1 c tomato juice
¼ c ketchup
½ c water

Combine ingredients; simmer 15 minutes or until slightly thickened.

Mrs. K. B., Cavalier Co., N. D.

BARBECUE SAUCE

1 pt. tomato puree or 1½ qts. juice
1 pt. vinegar
1 tbsp. cinnamon
1 tbsp. chili powder
1 tbsp. thyme
⅛ c Worcestershire sauce
2 tbsps. salt
½ tbsp. cloves
1 tbsp. dry mustard
2 tbsps. minced garlic
1½ c chopped onion
1 tbsp. allspice

Combine all ingredients; simmer slowly for 2 hours. May be sealed or kept covered in refrigerator indefinitely.

Mrs. G. M. E., Kingsbury Co., S. D.

BEEF MARINADE

½ c salad oil
¼ c vinegar
¼ c chopped onion
1 tsp. salt and pepper
2 tsps. Worcestershire sauce
2 lbs. round steak

Combine all ingredients except meat. Add meat; marinate for 1-3 hours. Grill until browned on both sides.

Mrs. J. S., Ward Co., N. D.

TOMATO SAUCE I

2 c canned tomato soup
1 tsp. chopped green pepper
1 tsp. chopped onion
1 tsp. butter or margarine
salt and pepper

Heat tomato soup to boiling; add onion, green pepper and butter. Season to taste. Simmer 5 minutes

E. H., Griggs Co., N. D.

TOMATO SAUCE II

2½ c fresh tomatoes, cubed or 2 c cooked or canned tomatoes
½ tsp. sugar (optional)
¼ c chopped onion
1 small bay leaf
2 cloves
2 tbsps. fat, melted
2 tbsps. flour
salt and pepper

Cook tomatoes, sugar, onion, bay leaf and cloves slowly for about 20 minutes for fresh tomatoes, 10 minutes for cooked or canned tomatoes. Put through a sieve or food press. Blend fat and flour; stir into tomato mixture. Cook over low heat stirring often until thickened. Season with salt and pepper. Makes 2 cups.

E. H., Griggs Co., N. D.

ALABAMA CHICKEN SPAGHETTI

- 1 (4-5 lb.) stewing chicken
- 3 celery stalks
- 2 tbsps. salt
- 4 c boiling water
- 6 medium onions, diced
- 1 c chopped celery
- 1 clove garlic, minced
- 2 (1-lb.) cans tomatoes
- 1 c ketchup
- 1 (6-oz.) can chopped mushrooms
- 3 tbsps. Worcestershire sauce
- 2 tsps. salt
- ½ tsp. pepper
- 1 (16-oz.) pkg. spaghetti

Place chicken in large kettle with celery stalks and 2 tablespoons salt. Add boiling water; simmer until tender (about 3 hours). Remove chicken from broth; cool; cut chicken from bones and dice. Skim ½ cup fat from cooled chicken broth; place in large kettle. Cook onions, celery and garlic in fat until onions are soft (not browned). Add chicken broth, tomatoes, ketchup, mushrooms and juice, Worcestershire sauce, salt and pepper. Bring to boil. Add spaghetti and cook until spaghetti is about half-cooked. Cool quickly. Makes 4½ quarts. To freeze, package in moisture-vapor-proof containers. Seal, label, date and freeze. (Recommended storage time: 3-6 months.) To serve, thaw in refreigerator for 8 hours or until mixture can be removed from container. Bake in covered casserole or baking pan at 400 degrees for about 45 minutes for 1 pint; 1 hour for a quart. If mixture appears dry, add 1/3 cup chicken broth or water before heating.

Variation: Cheese Chicken Spaghetti. Sprinkle chicken spaghetti with coarsely grated sharp cheese just before reheating.

S. C., Union Co., S. D.

CHICKEN-CORN SUPPER

- 1 c butter or margarine
- 2¼ c flour
- 2 qts. hot chicken broth
- 2 tsps. salt
- ½ tsp. pepper
- 2 tbsps. finely chopped onion
- 1¾ qts. chicken, chopped and cooked, or 2 stewing hens
- 1½ qts. cooked or canned whole kernel corn, or 3 (1-lb.) cans, drained
- 1½ c grated pasteurized process cheese
- ¾ c chopped pimientos
- buttered bread or cracker crumbs

Melt butter; blend in flour. Add hot broth; cook over medium heat until bubbling and thickened, stirring constantly. Stir in salt, pepper and onion. Combine with remaining ingredients, except crumbs, in large bowl or kettle. Cool quickly. Makes 4½ quarts. To freeze, pack in freezer containers, label, date and freeze. (Recommended storage time: 3-6 months.) To serve, thaw food enough to remove from container. Place in greased casserole; bake at 400 degrees for 30 minutes, or until thawed enough to press into the shape of baking dish. Sprinkle with buttered crumbs. Continue baking for about 1 hour for a pint, 1 hour 45 minutes for a quart, or until bubbly hot and crumbs are browned.

S. C., Union Co., S. D.

PIQUANT CHICKEN CASSEROLE

2 cans condensed cream of chicken soup
¼ c finely diced onion
2 tsps. salt
⅛ tsp. pepper
1 tsp. parsley flakes
1 tsp. celery flakes
¼ tsp. thyme
2 2/3 c water
2 2/3 c precooked rice
3 c diced cooked chicken
¼ tsp. paprika

Combine soup, onion and seasoning; mix well. Gradually add water, stirring constantly to blend thoroughly. Bring to a boil over medium heat, stirring constantly. Pour half the soup mixture into a 3-quart casserole. Then, making two layers of each, add the rice and the chicken. Top with remaining soup mixture. Cover and bake at 375 degrees for 15-20 minutes. Stir well after 10 minutes. Sprinkle with paprika.

Mrs. N. B. P., Codington Co., S. D.

CHICKEN SUPREME

1 can condensed celery soup
1 can condensed mushroom soup
¾ c dry rice
1 c milk
1 chicken, cut in pieces
1 pkg. onion soup mix

Place all ingredients in casserole, sprinkling onion soup mix over the top. Cover tightly. Bake at 350 degrees for 2½ hours.

Mrs. H. J., Sheridan Co., Mont.

FAR EAST CHICKEN CHOW MEIN

1 c chopped celery
½ c chopped onion
1 green pepper, cut in strips
5 tbsps. butter
¼ c flour
1 c evaporated milk
3 tbsps. soy sauce
2½ c (1-lb., 3-oz.can) bean sprouts, plus liquid
2 c chopped cooked chicken
2 tbsps. chopped pimiento
⅛ tsp. curry powder
5 c (two 5-oz. cans) Chinese noodles

In large saucepan saute celery, onion and green pepper in butter. Blend in flour; slowly add evaporated milk, stirring constantly. Cook until sauce is smooth and thickened. Add remaining ingredients, except Chinese noodles and cook until heated through. Serve over warmed Chinese noodles.

Ms. R. C., Wilkin Co., Minn.

TOPSY TURVEY TURKEY AND VEGETABLES

- 4½ c soft bread crumbs
- ¼ c diced celery
- ¼ c minced parsley
- ⅛ tsp. thyme
- ⅛ tsp. pepper
- 1/3 c water
- ¼ c butter
- 1 bouillon cube
- 2 c cubed, cooked turkey
- 2 c diced cooked vegetables
- 1 can condensed cream of celery soup

Combine bread crumbs, celery, parsley, thyme and pepper. Heat water with butter and bouillon cube until cube is dissolved; pour over crumb mixture and toss lightly. Pat half the crumb mixture onto sides and bottom of a greased 2-quart casserole. Mix turkey with vegetables and condensed celery soup; spoon into casserole. Top with remaining crumb mixture. Bake at 400 degrees for 35 minutes. Remove from oven, let stand 5 minutes then invert on serving platter.

Ms. P. M., Zieback Co., S. D.

BUTTER-BISCUIT TURKEY TWIST

- 2 c sifted flour
- 4 tsps. baking powder
- 1 tsp. salt
- 1 3 c butter
- 2/3 c milk
- 2 c diced cooked turkey
- turkey gravy

Sift together dry ingredients. Cut in butter until finely blended. Gently stir in milk. Turn out on floured board. Roll dough to 9x12-inch rectangle. Spread with chopped turkey. Beginning with the long edge, roll dough as for jelly roll. Shape into ring, sealed side down, on greased baking sheet. With scissors, cut to within ½-inch of center of ring at 1-inch intervals. Turn each section on its side. Bake at 425 degrees for 15-20 minutes. Serve hot with turkey gravy.

Ms. P. T., Moody Co., S. D.

TURKEY IN A SKILLET

- 2 tbsps. butter or margarine
- 2/3 c uncooked white rice
- 1 c whole kernel yellow corn, drained
- ½ c diced onion
- 2½ tsps. salt
- ¼ tsp. pepper
- 2 c diced, cooked turkey
- 1 (No. 2) can tomatoes
- 1 1/3 c water
- ¼ c chopped parsley

Melt butter or margarine in skillet. Add rice and cook, stirring occasionally until rice becomes yellow. Add corn, onions, salt, pepper, turkey and tomatoes. Slowly add water; mix well. Bring to a vigorous boil. Turn heat down. Cover and cook over low heat 14 minutes. Remove from heat, but leave lid on 10 minutes. Sprinkle with parsley just before serving.

Ms. Y. W., Roosevelt Co., Mont.

POULTRY/Turkey

BAKED TURKEY TEMPTER

- 2 tsps. salt
- 4 ozs. elbow macaroni or spaghetti
- ¼ c butter or margarine
- ¼ c flour
- 3 tbsps. nonfat dry milk solids
- ½ tsp. salt
- ⅛ tsp. pepper
- 1 c water
- 1¼ c (10½-oz. can) condensed cream of mushroom soup
- ¾ c cooked peas
- 1 c chopped cooked turkey (or chicken)
- ½ c (2 ozs.) shredded pasteurized process cheese

Add salt and spaghetti to 3 cups boiling water. Boil rapidly, stirring constantly, 2 minutes. Cover, remove from heat and let stand 10 minutes. Melt butter or margarine in saucepan. Stir in flour, milk solids, salt and pepper. Add water gradually, stirring constantly; blend in soup. Cook mixture until thickened, stirring constantly. Rinse macaroni or spaghetti with warm water; drain well. Add macaroni or spaghetti, peas and turkey to sauce. Turn into greased 1½-quart casserole. Sprinkle cheese over top. Bake at 350 degrees about 20 minutes.

Mrs. T. H., Burleigh Co., N. D.

TURKEY SQUARES

- 2 egg yolks, beaten
- 1½ c cooked white rice
- ½ c fresh bread crumbs
- 1 c cooked diced turkey
- ½ c finely chopped celery
- 1 c milk
- ¼ tsp. ginger
- ⅛ tsp. black pepper
- 1 tsp. salt
- 2 tbsps. chopped pimiento
- 2 egg whites, stiffly beaten

Combine all ingredients except egg whites. Fold in egg whites. Pour into heavily greased 7x11-inch pan. Bake at 350 degrees for 30-40 minutes, or until knife inserted in center comes out clean. Cut into squares and serve with sauce made from additional turkey or mushroom sauce.

Ms. D. M., Adams Co., N. D.

POULTRY/Stuffings

POULTRY STUFFING

- ½ loaf dry bread
- 3-4 c milk
- 1 egg
- 2 medium onions, diced
- ground giblets
- 1/3 c sugar
- salt and pepper

Soak bread in milk to make soft dressing. Add remaining ingredients; mix well. Stuff bird; close with skewers.

Mrs. M. B., Dunn Co., N. D.

CHICKEN DRESSING

- chicken giblets
- 1 tbsps. lard
- chopped onion
- salt
- pepper
- 2 c bread cubes
- ¼ c raisins, washed and drained
- 1 tbsp. sugar
- 1 tsp. baking powder
- ½ tsp. cinnamon
- ¼ tsp. salt
- 2-3 eggs

Fry giblets in lard until brown. Add chopped onion, salt and pepper; remove from heat. Combine all remaining ingredients except eggs; add to giblet mixture. Stir in eggs; mix well. Spoon into chicken cavity which has been seasoned with salt.

Mrs. E. R., Morton Co., N. D.

GOOSE STUFFING

- 1 c bread cubes per lb. bird
- 3 medium apples, peeled and sliced
- 1¼ c raisins, washed
- 1 large onion, cubed
- 1½ tsps. salt
- ¼ tsp. pepper
- ¼ tsp. sage
- ¼ tsp. thyme
- ¼ tsp. poultry seasoning
- giblets, cut-up
- 4½ c boiling water

Combine all ingredients. Pour boiling water over mixture; mix well; stuff goose. This recipe is for a 10-12-lb. bird.

Mrs. E. R., Morton Co., N. D.

STUFFING FOR FOWL

- 8 c dried bread cubes
- 6 medium potatoes, cooked and mashed
- 1 medium onion, chopped
- ¾ c chopped celery
- 3 eggs, well beaten
- ¼ c butter, melted
- salt
- pepper
- parsley

Combine all ingredients. Stuff bird or bake as a casserole.

Mrs. O. M. A., Bottineau Co., N. D.

MEATS & POULTRY, ROASTING CHART

Cut	*Weight in lbs.*	*Oven Temp.*	*Internal Temp.*	*Cooking Time*
Beef				
Standing Rib	4-6	325	140 (rare)	2¼-2¾ hrs.
			160 (medium)	2¾-3¼ hrs.
			170 (well)	3¼-3½ hrs.
Rolled Rib	5-7	325	140 (rare)	3¼-3½ hrs.
			160 (medium)	3¾-4 hrs.
			170 (well)	4½-4¾ hrs.
Rolled Rump	4-6	325	150-170	2-2½ hrs.
Sirloin Tip	3½-4	325	150-170	2-2¾ hrs.
Rib Eye or Delmonica	4-6	350	140 (rare)	1½-1¾ hrs.
			160 (medium)	1¾ hrs.
			170 (well)	2 hrs.
Tenderloin, whole	4-6	425	140 (rare)	45 min.-1 hr.
Tenderloin, half	2-3	425	140 (rare)	45-50 min.
Veal				
Leg	5-8	325	170	2¾-3¾ hrs.
Loin	4-6	325	170	2½-3 hrs.
Rolled Shoulder	4-6	325	170	3½-3¾ hrs.
Pork				
Loin, center	3-5	325	170	2½-3 hrs.
Loin, half	5-7	325	170	3½-4¼ hrs.
Loin, blade	3-4	325	170	2¼-2¾ hrs.
Loin, center; rolled	3-4	325	170	2½-3 hrs.
Boston Shoulder	4-6	325	185	3½-4½ hrs.
Boston Shoulder Roll	3-5	325	185	3-3½ hrs.
Leg	10-14	325	185	5½-6½ hrs.
Leg, half	5-7	325	170	3½-4½ hrs.
Lamb				
Leg, whole	5-8	325	175-180	3½-3¾ hrs.
Leg, half	3-4	325	175-180	3-3½ hrs.
Square Cut Shoulder	4-6	325	175-180	2½-3 hrs.
Rolled Shoulder	3-5	325	175-180	2¾-3 hrs.

FISH & GAME

The Dakotas have perhaps the nation's most abundant and varied supply of wild game and fish. Indians and pioneers who first inhabited the Dakota prairies relied on this natural meat supply for a portion of their livlihood. Dakotans today continue to include wild game and fish in their menues, and Dakota housewives are adept at wild game cookery. In this section you will find their favorite recipes for preparing pleasing mail dishes from the bounty of stream and field.

Buffalo
Duck
Goose
Grouse
Pheasant
Venison
Carp
Perch
Pike
Trout

buffalo

Buffalo are synonymous with the "Old West." They provided the backbone of the Indian's livelihood. But with the coming of white men, the buffalo faced extinction as millions were slaughtered by hide hunters. In 1900 only 6 wild animals and a few hundred in private herds remained of the estimated 60 to 125 million that were here when Columbus came. Today a small herd roams under government protection in Custer State Park and a growing number of ranchers are raising buffalo for meat. Buffalo meat is unusually low in cholesterol and saturated fats and high in protein and polyunsaturates. Those that have access to buffalo meat have found it is delicious.

BUFFALO ROAST

1 (7-lb.) rib eye buffalo roast
Marinade Sauce:
2 c hard cider
2 c fresh sweet cider
½ c red wine vinegar
2 onions, chopped
1 carrot, chopped
3 sprigs parsley
¼ tsp. crumbled dried thyme
1 bay leaf, crumbled
6 whole juniper berries
10 whole black peppercorns

Combine marinade ingredients. Bring to boil over high heat; reduce heat and simmer partially covered for 30 minutes. Pour into crock or casserole; cool to room temperature. Add meat to marinade, turning to coat all sides. Add enough water so marinade reaches half way up meat. Cover; marinate at room temperature for at least 24 hours, turning every few hours. Dry meat; place on rack in shallow roasting pan. Strain marinade through fine sieve, reserving liquid. Roast meat at 500 degrees for 20 minutes. Reduce heat to 350 degrees; baste with ½ cup marinade. Continue roasting, basting every 15 minutes. Cook as for beef roast.

Mrs. L. N., Custer Co., S. D.

DEEP FAT FRIED PHEASANT

1 young pheasant, cut in pieces
¼ c coating mixture
milk or buttermilk
cooking oil

Marinate pheasant pieces in milk or buttermilk 1-2 hours in refreigerator. Dredge pieces in desired coating mixture. Dry on rack about ½ hour. Transfer a few coated pieces at a time to 340-360 degree fat. Remove pieces when golden brown (3-5 minutes).

Coating mixtures:

(1) Mix together ¼ cup flour, 1 teaspoon paprika, ¾ teaspoon salt, and ⅛ teaspoon pepper.

(2) ¼ cup pancake mix and ¼ teaspoon salt.

(3) Mix together ¼ cup flour, ¾ teaspoon salt, ¼ teaspoon pepper, 1/16 teaspoon oregano, 1/16 teaspoon basil.

(4) Beat until smooth 1 egg, ½ cup milk, ½ cup flour, ½-1 teaspoon Worcestershire sauce, ⅛ teaspoon allspice, ¼ teaspoon salt and ⅛ teaspoon pepper.

Mrs. A. C., Logan Co., N. D.

ROAST PHEASANT

1 young pheasant
salt
dressing (optional)
8 slices salt pork or bacon
¼ c oil

Rub cavity of pheasant with salt. Stuff with favorite dressing if desired. Completely cover breast and all meaty portions with strips of salt pork or bacon; tie in place. Place bird breast side up on rack in shallow roasting pan. Pour oil over bird. Bake at 400 degrees, uncovered, for about 50-60 minutes, basting with oil and turning occasionally.

Mrs. W. A., Brown Co., S. D.

PHEASANT OVER CHARCOAL

2 pheasants, cut in pieces
Marinade Sauce:
¼ c cooking oil
2 tbsps. chopped onion
salt and pepper
garlic
¼ c white wine
2 tbsps. vinegar
2 tbsps. Worcestershire sauce
1 tsp. sugar
1 tsp. Tabasco sauce
1 tsp. paprika

Mix together all sauce ingredients. Turn pieces of meat in sauce; place meat on charcoal grill. Brush frequently with marinade sauce; cook until deep brown and well done.

Mrs. W. J., Lake Co., S. D.

PHEASANT IN CREAM SAUCE

1 (1½-2 lb.)) pheasant, cut up
2 c boiling water
1 medium onion, sliced
2 stalks celery, with tops
1 tsp. salt
⅛ tsp. pepper
1 c light cream
2 tbsps. flour
2 tbsps. cold water
2-3 tbsps. red wine

Cover pheasant with boiling water; add onion, celery, salt and pepper. Simmer, covered, for about 45 minutes, or until pheasant is tender. Remove pheasant from broth; take meat off bones. Strain broth; boil down to ½ cup. Add cream; heat to boiling. Stir in blended flour and cold water; cook, stirring, until mixture thickens. Add pheasant; heat thoroughly. Add red wine; serve.

Mrs. J. W., Burleigh Co., N. D.

PHEASANT POT PIE

1 pheasant, dressed and cut into pieces
1 small onion, sliced
1 carrot, sliced
1 bay leaf
3 peppercorns
½ tsp. salt
1 box frozen peas, cooked
1 c white sauce
1 recipe baking powder biscuit dough

Simmer pheasant in salted water with onion, carrot, bay leaf and peppercorns until tender and meat falls from bones. Remove meat from bones; place chunks of meat in well-buttered casserole. Add cooked peas. Cover with white sauce. Cut biscuits from dough; cover contents of casserole. Bake at 450 degrees for 15 minutes, or until biscuits are done.

Mrs. D. C., Spink Co., S. D.

pheasants

The pheasant is a symbol of South Dakota. We have one-half million of these beautiful birds coloring our countrysides, roosting in our shelterbelts, and, during the hunting season, serving as a seasonal delicacy on our dinner tables.

SCALLOPED PHEASANT

4 c dry bread cubes
2 eggs, slightly beaten
2 c chopped, cooked celery
1 small onion, grated
½ tsp. sage
salt and pepper
4 c cooked, chopped pheasant

Custard:
1 c fat from stock or butter
1 c flour
1 qt. stock
1 qt. milk
6 eggs, beaten

Mix together bread, eggs, celery, onion, sage and seasonings. Put a layer in large, well-greased baking dish; add a layer of pheasant. Continue layering until all ingredients are used. Add custard, moving ingredients slightly with fork so custard runs to bottom of pan. Bake at 325 degrees for 1 hour and 15 minutes.

Custard: Combine fat or butter, flour, stock and milk; cook until thick. Remove from heat; add beaten eggs.

Mrs. G. T., Campbell Co., S. D.

PHEASANT AND APPLES WITH SOUR CREAM SAUCE

2 (1-lb.) oven-ready pheasants, split lengthwise into halves
1 tsp. salt
¼ tsp. ground black pepper
½ c flour
4 slices bacon
3 tbsps. butter
½ c finely chopped onions
pheasant giblets, finely chopped
1 c chicken stock, fresh or canned
2 unpeeled cooking apples, cored and quartered
2 tbsps. brandy
2 tsps. flour
½ c sour cream

Sprinkle pheasant halves with salt and pepper. Dip each half in ½ cup flour, coating both sides; shake off excess flour. Fry bacon over moderate heat until crisp; drain on towel and crumble into bits. Drail all but 2 tablespoons fat from skillet; add butter; brown pheasant halves evenly; set aside on plate. Add onions to remaining fat; cook about 5 minutes, until soft and translucent. Add giblets, stir for 2-3 minutes. Stir in stock; return birds; bring to boil over high heat; reduce heat; simmer partially covered for 20 minutes. Add apple quarters, coating with liquid. Simmer partially covered for about 10 minures longer. (Birds are done when juice from punctuured thigh is yellow.) Arrange pheasants and apples on heated platter; cover with foil. Skim fat from surface of cooking liquid; pour in brandy; bring to simmer over moderate heat. Reduce heat; cook for 1-2 minutes. Beat 2 teaspoons flour into sour cream; add to sauce. Simmer slowly for 4-5 minutes until smooth and slightly thickened. Pour sauce over pheasants; sprinkle with bacon bits; serve immediately.

Mrs. L. N., Walsh Co., N. D.

game birds & animals

The many game birds and animals of the Dakotas provide good eating and hours of sport for many Dakotans. Small game seasons are open on grouse, partridge, pheasant, wild turkey and quail. Waterfowl hunting includes snipe, geese, duck and sandhill crane, and big game hunters bag deer, antelope and Black Hills elk.

WILD DUCK

- 2 ducks
- 3 tbsps. cooking oil
- 1 large onion, chopped
- 3 large carrots, chopped
- 5 stalks celery, chopped
- 1 c red wine
- 1 can beef consomme, undiluted
- 2 tbsps. currant jelly
- salt and pepper
- 2 tbsps. cornstarch
- lemon juice or brandy (optional)

Put ducks in oil in pan. Bake at 400 degrees for about 30 minutes, turning occasionally until brown all over. In another pan combine onions, carrots and celery. Lay birds on top. Bake at 275-300 degrees for 1 hour longer. Baste several times with water, turning birds over each time. Remove birds from pan. Add wine, consomme, jelly, salt and pepper to vegetables. Cook on top of stove for 10 minutes. Strain. Thicken by pouring off small amount of liquid into cup and adding cornstarch. Mix well; return to pan. Cook a few minutes longer. Add lemon juice or brandy at last minute.

Mrs. W. C., Spink Co., S. D.

WILD DUCK IN WINE SAUCE

- 2 wild ducks, halved or quartered
- 2 tbsps. margarine
- 2 tbsps. flour
- 1 c chicken broth
- ½ c burgundy wine
- 2 tbsps. chopped onion
- 1 small bay leaf
- ½ tsp. salt
- dash of pepper
- 1 (3-oz.) can mushrooms

Simmer ducks covered in small amount of salted water for 20-30 minutes. Drain. In skillet brown ducks in margarine. Remove ducks. Blend flour in margarine in skillet. Add chicken broth, wine, onion, bay leaf, salt and pepper. Cook and stir until bubbly. Add mushrooms. Pour over duck which has been placed in roasting pan. Cover and bake at 350 degrees for 1¼ to 1½ hours, or until tender. Transfer duck to platter. Remove bay leaf; skim excess fat from sauce. Pass with duck.

Mrs. D. L., Brown Co., S. D.

WILD GAME/Grouse

RUFFED GROUSE WITH BRANDIED ORANGE STUFFING

- 3 ruffed grouse
- livers from grouse, chopped
- 2¼ c prepared sage and onion stuffing
- ¼ lb. butter
- 6 tbsps. brandy
- 1½ tbsps. grated orange rind
- ½ c orange pulp
- 1 c diced celery
- 1 egg, beaten
- pinch of rosemary
- pinch of marjoram
- pinch of chervil
- salt and pepper
- French dressing (3 parts oil to 1 part wine vinegar or lemon juice)

Saute stuffing and livers in 2 tablespoons butter. Mix in brandy, orange rind and pulp. Set aside. Saute celery in 2 tablespoons butter; add remaining butter (melted), eggs, seasonings, salt and pepper. Blend thoroughly with stuffing mixture. Stuff and truss grouse. Cook bird on rotisserie, brushing with French dressing every 15 minutes until done. Remove; dash birds with brandy.

Mrs. G. E. S., Sheridan Co., N. D.

WILD GAME/Goose

ROAST GOOSE

- 1 wild goose
- juice of 1 lemon
- 3 c soft bread crumbs
- salt and pepper
- ¼ c butter, melted
- ¼ c chopped onion
- 1 c chopped apple
- 1 c chopped dried apricots
- 3 c soft bread crumbs
- 4 slices bacon

Sprinkle goose inside and out with lemon juice, salt and pepper. Combine melted butter and onion in large saucepan; cook until tender. Stir in apple, apricots, bread, salt and pepper. Stuff goose. Close opening with skewers; string. Cover breast of goose with bacon slices. Place breast side up on rack in roaster. Roast at 325 for 20-25 minutes per pound, or until tender and done. Baste frequently with bacon fat and pan drippings.

Mrs. T. R., Walworth Co., S. D.

ROAST LEG OF VENISON

2 tbsps. butter
1½ c finely chopped onions
1 carrot, scraped and finely chopped
¼ c finely chopped celery
1 tsp. finely chopped garlic
3 c dry red wine
1 c cold water
1 bay leaf, crumbled
6 whole juniper berries and 10 whole black peppercorns, wrapped in towel and crushed with rolling pin
1 tbsp. salt
1 (10-11-lb.) leg venison
1 tbsp. cornstarch
1 tbsp. cold water

Melt butter over moderate heat; add onions, carrot, celery and garlic; cook for about 5 minutes or until vegetables are soft but not brown. Stir in wine, 1 cup water, bay leaf, juniper berries and peppercorns, and salt; bring to boil over high heat. Remove pan from heat; let cool to room temperature. Place venison in roasting pan; pour marinade over it; turn meat to moisten all sides. Cover with foil; refrigerate at least 12 hours, turning 2-3 times. Roast venison at 425 degrees for 15 minutes. Reduce heat to 375; roast for 1 hour and 15 minutes longer, basting every 15 minutes with pan liquids. (Meat is done when meat thermometer reads 140-150 degrees.) Prepare sauce by straining cooking liquid through fine sieve. Skim fat from surface. Sauce should measure 2 cups. Bring to simmer over moderate heat, add mixture of cornstarch and 1 tablespoon cold water, stirring constantly with wire whisk. Cook until slightly thickened and smooth. Serve at once with roast venison.

Mrs. J. Z., Kingsbury Co., S. D.

BRAISED LEG OF VENISON

1 (9-10-lb.) leg of venison
larding pork
Marinade:
1 pt. wine vinegar
1 small onion
1 carrot, sliced
1 stalk celery
pickling spices
1 qt. buttermilk
salt and pepper
beef or veal stock
1 c sour cream
1 tbsp. red currant jelly
few drops lemon juice

Coat venison leg with lard on all sides. Combine first 5 ingredients of marinade; boil for 15 minutes; cool. Add buttermilk and leg; let stand 4-5 days. Take out leg; rinse with water; dry. Season to taste. Roast at 350-375 degrees, adding some of vegetables from marinade and stock, until tender (about 15 minutes per pound). Baste frequently. When done, remove leg and keep warm. Add flour to thicken pan juices; add remaining ingredients. Strain; serve with ¼-inch thick slices of leg.

M. C. C., Milwaukee, Wis.

ELK MEATBALLS IN SOUR CREAM

6 tbsps. butter
½ c finely chopped onions
1 tsp. finely chopped garlic
2 lbs. ground elk or venison
1 lb. lean ground pork
1 c soft fresh bread crumbs
1 egg, slightly beaten
½ c milk
¼ c finely chopped parsley
1 tsp. crumbled dried thyme
1 tbsp. salt
¼ tsp. ground black pepper
2 tbsps. vegetable oil
2 c chicken stock
1 c sour cream
2 tbsps. flour
2 tsps. dry mustard

Melt 2 tablespoons butter over moderate heat. Add onions and garlic; cook about 5 minutes, stirring frequently, until soft and translucent but not brown. Add to elk, pork, bread crumbs, egg, milk, parsley, thyme, salt and pepper; knead vigorously. Beat mixture with wooden spoon until smooth and fluffy. Roll into 1-inch meatballs. Melt 4 tablespoons butter with oil; brown meatballs, 10-12 at a time. Remove meatballs; drain off fat. Bring chicken stock to boil in skillet over high heat. Return meatballs to skillet, turning to moisten evenly. Reduce heat;, simmer partially covered for 20 minutes, or until browned through. Transfer meatballs to heated platter; prepare sour cream sauce. Beat sour cream, flour and mustard until well blended. Add to liquid remaining in skillet; cook over low heat for 4-5 minutes, stirring constantly until thickened. Pour sauce over meatballs. Serve immediately with hot buttered noodles or rice.

Mrs. L. P., Walsh Co., N. D.

FISH

Fish	Cooking Method
Blue Pike	Bake, boil, fry
Brook Trout	Bake, boil
Carp	Fry, bake, boil
Catfish (Bullhead)	Fry, boil, broil
Lake Trout	Bake, broil, saute
Perch	Broil, bake, fry
Pike	Broil, bake, fry
Walleye	Broil, bake fry

facts about fish

*Fish should be dressed as soon as they are taken ashore
*Protect fish from exposure to sun and wind
*When fish is dressed, pack in wet grass or ice until cooking

FRYING: Use oil or drippings. Rinse fish in cold water. Drop into mixture of heavily salted flour, corn meal or bread crumbs. Allow 4 teaspoons salt for ½ cup coating. Put into hot grease. Brown 3 minutes . . . take off heat for 2 minutes . . . return to heat for 3 minutes more on other side.

DEEP FAT FRYING: Dip fish in well-salted flour, beaten egg or undiluted evaporated milk. Dip in fine dry bread crumbs. Use frying basket. Put single layer of fish in frying basket. Fry 4-6 minutes, or until golden brown.

STEAMING FISH: Wrap fish in cheesecloth or place in wire basket with clove of garlic. Use pan with rack. Put 1½ inches of water in bottom of pan. Lay fish on rack; cover pan. Boil water, allowing steam to cook fish (about 15 minutes per pound).

BAKING FISH: Either whole fish, fish steaks, or fillets may be baked. Dry fish require frequent basting, or lay strips of salt pork or bacon over surface to prevent drying out. Both steaks and fillets may be dipped into slightly salted milk and rolled in soft crumbs before baking. With large fish bake in very hot oven for first 15 minutes, reducing then to 375-400 degrees for remainder of cooking time.

BOILING FISH: Place fish on perforated pie plate; tie in cheesecloth; immerse in salted water just below boiling point. Bring quickly to boiling point, then reduce heat to avoid breaking tender flesh. Allow 6-10 minutes cooking time per pound. To insure firmness of texture, 1 tablespoon vinegar or lemon juice may be added to cooking water. For flavor, add onion slice, 1-2 cloves, bay leaf, and sprig of thyme or basil.

BROILING FISH: Grease broiler pan; lay on fish. Place close to heating unit until top surface is seared; turn; sear second side. Move further from heating unit to complete cooking, basting frequently with melted butter for dry fish.

BAKED CARP WITH VEGETABLES

3 lbs. carp
2 c diced celery
2 c diced onions
salt
pepper
1 clove garlic (optional)
1 c hot water
½ c cooking oil or drippings
minced parsley

Clean, skin and fillet fish; cut into chunky pieces. Combine celery, onions, salt and pepper; arrange in greased baking dish, burying garlic in them if desired. Place fish on top of vegetable; pour on oil or drippings and water; cover closely. Bake at 350 degrees for 1½ hours. Serve topped with parsley.

Mrs. J. K., La Moure Co., N. D.

BAKED CARP

3 lbs. carp
1 tbsp. salt
1 tsp. pepper
½ c sliced onion
1 pt. sour cream
1 tbsp. paprika
¼ c margarine

Clean and skin fish. Wash thoroughly and dry. Season inside and out with salt and pepper. Place in buttered baking pan; pile sliced onions on top. Cover with blended sour cream and paprika. Bake one hour, or until flesh loosens from bones. Baste with melted margarine during baking.

Mrs. E. K., Burleigh Co., N. D.

SPICED CARP

3 lbs. carp
2 c vinegar
1 bay leaf
2 tbsps. pickling spices
2 onions, thinly sliced
1 lemon, thinly sliced

Clean, skin, fillet and cut fish into chunky pieces; steam or boil until just tender (about 10 minutes). Lift carefully from pan; place in dish. Scald together vinegar, bay leaf and pickling spices. Pour over fish. Arrange onions and lemon over fish; cover; let stand in refrigerator overnight. Turn fish in vinegar 2 or 3 times befor e using.

Mrs. M. J., Mercer Co., N. D.

PAN FRIED PERCH

1½ lbs. perch
¼ c salad dressing
½ c frying fat
½ c corn meal
1 tsp. salt
½ tsp. pepper

Dip fish in salad dressing until lightly coated. Dip in mixture of corn meal, salt and pepper. Fry in hot fat until brown.

Mrs. L. M., Charles Mix Co., S. D.

FLAKED PERCH

½ c frying fat
4 lbs. perch
2 eggs, beaten
2 tsps. salt
½ tsp. pepper
4 tbsps. grated cheese
1 c crushed corn flakes

Wash, scale and clean fish. Combine eggs, salt, pepper and grated cheese; beat well. Dip each fish in this mixture; roll in corn flakes. Fry fish in hot oil for five minutes or until brown.

Mrs. H. B., Dickey Co., N. D.

Pike/**FISH**

SPINNING REEL PIKE

4 lbs. pike
1 tbsp. salt
1 tbsp. water
1 egg, beaten
1½ c cracker crumbs
2 slices salt pork, cut small
2 c thin cream
3 c onion
2 tbsps. olive oil
1 tsp. sugar

Remove head, clean and rinse the fish. Rub inside with salt. Combine egg and water; brush over outside of fish. Dredge fish in cracker crumbs. Place on greased baking pan. Slash across backs of fish; insert bits of salt pork. Bake at 425 degrees for 40 minutes. Baste often with cream. While fish bakes, cook onions in salted water until tender. Remove and brown in olive oil and sugar mixture. Serve baked fish on platter; pour on onion sauce.

Mrs. C. C., Campbell Co., S. D.

POACHED PIKE

3 lbs. pike
1 bay leaf
1 tbsp. salt
2 slices onion
½ tsp. celery salt
¼ tsp. pepper
1 tbsp. lemon juice
4 tbsps. butter
3 tbsps. flour
1 egg, slightly beaten
2 tbsps. cream

Fillet the fish. Fill large skillet half full of water; add bay leaf, salt, onion, celery salt, pepper and lemon juice; boil 5 minutes. Add fish; simmer 10 minutes. Remove fish to platter. The liquid left is the fish stock. Melt butter; blend in flour until it makes smooth paste. Add 1¾ cups fish stock. Stir constantly over low heat until mixture thickens. Mix together egg and cream; add to stock mixture, stirring constantly until sauce thickens. Pour over fish.

Mrs. B. O., Sheridan Co., N. D.

BAKED PIKE WITH DILL SAUCE

4 lbs. pike
1 tbsp. salt
1 egg, slightly beaten
1 tbsp. water
fine dry bread crumbs
1 thick slice salt pork, cut in 2-in. strips
2 c thin cream
3 c small pickling onions
2 tbsps. butter
1 tsp. sugar

Dill Sauce:
1 egg yolk, hard-cooked
1 egg yolk
½ c salad oil
¾ tbsp. vinegar
⅛ tsp. Worcestershire sauce
½ tsp. dry mustard
¼ tsp. salt
¼ tsp. sugar
1/16 tsp. pepper
¼ c heavy cream, whipped
½ tsp. dill seed

Wash fish; dry. Rub inside with salt; brush outside with egg and water mixture. Roll in crumbs. Place in baking pan, side resting on bottom of pan. Arrange pork strips in shallow slashes across back of fish. Bake at 425 degrees for 40 minutes. When fish has begun to brown, baste with cream every 5 minutes. Cover onions with boiling salted water; cook 10 minutes. Drain; cook in butter and sugar until tender and browned. Garnish fish with onions and serve with dill sauce.

Dill Sauce: Rub cooked egg yolk through sieve; add uncooked egg yolk. Beat in salad oil gradually; add vinegar, Worcestershire sauce and seasonings. Just before serving fold in whipped cream and dill seed.

Mrs. B. S., Potter Co., S. D.

DRESSED UP PIKE

- 6 lbs. pike
- 2 loaves bread
- water
- 3 tsps. salt
- 1 tsp. pepper
- 2 tsps. poultry seasoning
- ½ onion, diced
- 1 onion, sliced
- 1 c tomato juice

Clean fish. Make stuffing by softening bread with water until of proper consistency. Season with salt, pepper, poultry seasoning and ½ diced onion. Stuff fish. Rub baking pan with grease; lay on fish. Bake at 350 degrees until fish begins to brown; add 1 sliced onion and tomato juice. Bake 1½ hours longer. To serve, cut down backbone and portions along the side. Handle carefully to get meat off bones.

Mrs. S. F., Walworth Co., S. D.

Trout/**FISH**

FALLING LEAF TROUT

- 4 lbs. trout
- ½ c flour
- 1 tsp. salt
- ½ tsp. pepper
- 1¼ c cooking oil
- 3 onions, chopped
- 1 clove garlic, diced
- 1 (No. 2) can tomatoes
- 1 small can tomato paste
- ½ tsp. sugar
- 3 green peppers, sliced

Wash, scale and clean fish. Remove from bone. Shake fillets in paper bag with flour, salt and pepper. Fry fish in 1 cup oil. Drain on paper towel. For sauce, saute onions in ¼ cup oil. Add remaining ingredients, except green peppers. Simmer for 1 hour. Add green peppers; simmer until peppers are tender. Pour over fried fish; serve.

Ms. O. K., Barnes Co., N. D.

SPICED LAKE TROUT

- 1 large onion, diced
- 2 lbs. lake trout
- 2 qts. water
- ½ c vinegar
- 2 tbsps. mixed spices
- 1 c brown sugar
- 1 tsp. salt
- 14 ginger snaps

Add onions and fish to water; boil slowly until fish is cooked (about 15 minutes). Remove fish. Add vinegar, spices, brown sugar and salt to water. Thicken with ginger snaps. Pour over fish and serve.

Mrs. J. T., Brown Co., S. D.

FISH/Trout

BAKED LAKE TROUT

- 2 lbs. trout
- 1 onion, sliced
- 2 tsps. Worcestershire sauce
- 2 tbsps. butter, melted
- 1 lemon, sliced
- 1 tsp. salt
- ½ tsp. pepper
- lettuce

Clean trout; wipe dry. Sprinkle onion slices with Worcestershire sauce; season with salt and pepper. Line baking pan with leaf lettuce; lay fish on top. Brush with butter; top with thin slices of lemon. Bake at 350 degrees for about 45 minutes, or until fish loosens from bones when touched with fork. Remove lettuce and lemon before serving.

Mrs. G. H., McLean Co., N. D.

FISH/Seafood

TUNA WIGGLE

- 1 small can peas
- 2 cans tuna, flaked
- 1 green pepper, minced
- 2 tbsps. flour
- 1 tbsp. butter
- 2 c milk
- seasoning to taste

Melt butter; blend in flour and seasonings. Add milk and allow to thicken. Add remaining ingredients. Heat and serve hot.

H. B., Walsh Co., N. D.

QUICK TUNA HOT DISH

- 1 can tuna
- 1 (1½-lb.) bag potato chips
- 1 can mushroom soup
- 1 soup can of milk

Spread half of tuna in bottom of casserole; crumble half of potato chips over tuna. Mix soup and milk together; pour half of potato chips over tuna. Repeat layers. Bake at 350 degrees for 25 minutes.

Mrs. H. J., Sheridan Co., Mont.

EGG PLANT CASSEROLE

- 1 medium-sized eggplant
- 1 (7-oz.) can tuna
- 1½ c medium white sauce
- ½ c buttered bread crumbs

Peel eggplant and cut in ¾-inch cubes. Cook until tender in salted water. Drain at once. Place eggplant in well-greased baking dish; spread tuna evenly over it. Pour in white sauce, working it into layers with back of spoon. Top with bread crumbs. Bake at 375 degrees for 20-30 minutes, or until lightly browned.

Mrs. A. V., Hand Co., S. D.

SALMON LOAF

2 c (1-lb. can) salmon
2 eggs, slightly beaten
½ c milk
½ tsp. salt
⅛ tsp. pepper
1 tsp. finely chopped onion
1 tbsp. finely chopped green pepper
1 tsp. lemon juice
3 c Special K cereal

Remove bones and skin from salmon; drain and flake. Add eggs to salmon; mix well. Stir in remaining ingredients; spread in well-greased 9½x5¼-inch loaf pan. Bake at 350 degrees for 45 minutes. Serve hot.

Mrs. P. G., Roberts Co., S. D.

SHRIMP CASSEROLE

2 tbsps. butter
1 small onion, finely chopped
3 c cooked rice
1 small green pepper, sliced
1 (4-oz.) can button mushrooms
1 (8-oz.) can tomato sauce
1½ tsps. salt
½ tsp. black pepper
1 lb. shrimp

Saute onion in butter in preheated skillet. Add remaining ingredients; mix well. Cover and simmer for about 40 minutes.

Mrs. F. H., Emmons Co., N. D.

Miscellaneous/**FISH**

PICKLED FISH

1 qt. fish chunks
5/8 c salt
white vinegar
2 c sugar
1 c white vinegar
2 medium onions, chopped
¾ box mixed spices

Add salt to fish; cover with white vinegar. Let stand 5 days. Drain; rinse with water; soak ½ hour in water and drain. Mix together sugar, 1 cup white vinegar, onions and spices; pour over fish. Let stand 3 days.

Mrs. L. G., Aurora Co., S. D.

SOUP 'N SANDWICH

There are some days when nothing sounds quite as good as a piping hot bowl of homemade soup. Even back in the pioneering days, one of the favorite ways to use up those fresh garden vegetables was in soup, both hearty and nutritious. And what goes better with soup than a sandwich? If today is one of those soup 'n sandwich days, we invite you to try one of the combinations you'll find here.

Chowder
Soups

Cold Sandwiches
Hot Sandwiches

CHILI SOUP

2 lbs. hamburger
1½ large onions, chopped fine
1 clove garlic, minced
2 (No. 2) cans kidney beans
2 qts. tomato juice
2 tbsps. flour
1 tbsp. sugar or honey
1½ tbsps. chili powder
1 scant tbsp. paprika
2 tsps. salt

Brown hamburger, onions and garlic. Add kidney beans and tomato juice; simmer. Mix together remaining ingredients; add to soup. Simmer 1 hour. Stir occasionally.

Mrs. A. A. B., Hutchinson Co., S. D.

FRENCH ONION SOUP

6 medium onions, sliced
¼ c butter or margarine
5 bouillon cubes, dissolved in 1 qt. boiling water
1 tsp. Worcestershire sauce
½ tsp. salt
½ tsp. paprika
6 slices French bread, toasted
½ c grated Parmesan cheese

Cook onions in butter until straw-colored. Add dissolved bouillon cubes and seasonings. Simmer, covered, until onions are tender. Pour soup into earthenware soup dish; place toast on top. Sprinkle with cheese. Bake in hot oven until cheese melts.

Mrs. C. W., Cass Co., N. D.

FRANKFURTER AND POTATO SOUP

2 c diced potatoes
1 small onion, diced
1½ c boiling water
2 frankfurters, cut in ¼-inch slices
1¾ tsps. salt
pepper
2 c milk
2 tbsps. finely chopped parsley

Cook potatoes and onion in boiling water until soft. Put through a ricer or mash slightly. Add frankfurters, seasonings and milk to potato mixture. Heat thoroughly; add parsley and serve.

Variation: Salami or other luncheon meat may be substituted for the frankfurters. Fresh sausage may be used if it is fried until crisp and drained before adding it to the soup.

Mrs. A. V., Beadle Co., S. D.

HAM-SPLIT PEA SOUP

1 c dried split peas
3½ qts. water
1 small ham hock
½ c chopped onion
3 tbsps. butter
3 tbsps. flour
2 c milk
½ tsp. salt
⅛ tsp. pepper

Soak split peas overnight in water; add ham hock and onion. Cook until ham is falling off bone. Remove hock; cut ham in small pieces. Combine remaining ingredients; add to pea mixture; cook until thick.

Mrs. R. E., Ill.

HEARTY VEGETABLE-BEEF SOUP

1 meaty soup bone (beef or knuckle of veal)
3 qts. water
1 c chopped onion
2 c chopped celery
2 c tomatoes
2 c sliced carrots
2 c diced turnips
salt
pepper

Simmer soup bone in water for 4 hours, or until meat is tender. Remove soup bone; cut meat from bone. Put part of meat into broth; reserve rest for hash or croquettes. Add remaining ingredients to broth; simmer until vegetables are tender.

Mrs. M. L. C., Mo.

PRESSURE COOKER VEGETABLE SOUP

2 lbs. soup meat
4 c water
½ c diced carrots
¼ c diced potatoes
½ c chopped celery and leaves
¼ c sliced onions
¼ head cabbage, chopped
2 c canned tomatoes
½ c rice
1 tbsp. salt
¼ tsp. pepper

Combine all ingredients in cooker. Cook for 30 minutes; let pressure go down; open and serve.

Mrs. E. A., Butte Co., S. D.

POTATO SOUP

5 strips bacon
5 medium-sized potatoes, chopped fine
5 onions, chopped fine
3 tsps. salt
1 c water
5 c milk

Cut bacon into ½-inch pieces; fry slowly in deep kettle. Add potatoes, onions, salt and water. Cook slowly so no extra cooking water will be needed. When vegetables are done, mash without draining. Add milk and bring to boil.

Mrs. M. L. J., Walsh Co., N. D.

TOMATO SOUP

23 lbs. ripe tomatoes	3 green peppers
4 stalks celery	3 bay leaves
14 pieces parsley or parsley flakes	2 c flour
	2 c brown sugar
10 medium-sized onions	scant ½ c salt
30 whole cloves	1 lb. butter

Boil together until soft tomatoes, celery, parsley, onions, cloves, peppers and bay leaves. Put mixture through a sieve. Make cream sauce of remaining ingredients; add to tomato mixture; boil until thickened. Pour into pint jars; seal.

Soup: to make soup add 1 pint tomato soup to 1 pint hot milk.

Mrs. L. M., Haakon Co., S. D.

TOMATO BISQUE

1 pt. tomatoes, sieved	1/16 tsp. pepper
2 tbsps. minute tapioca	3 tsps. sugar
2 tbsps. butter	⅛ tsp. soda
1½ tsps. salt	1½ pts. hot milk or cream

Heat tomatoes in double boiler; add tapioca, butter, salt, pepper, sugar and soda. Cook 15 minutes, stirring frequently. When ready to serve gradually stir tomato mixture into hot milk or cream. (Never stir milk into tomatoes, as it will curdle.) Serve hot.

Mrs. A. F., Ward Co., N. D.

HAM & BEAN SOUP

1 c navy beans	½ c chopped onion
1 small meaty ham bone	3½ c canned tomatoes
1½ qts. water	1 tbsp. minced parsley
1 c diced celery	salt and pepper

Soak beans overnight in cold water; drain. Cover beans and ham bone with water; add celery, onion, tomatoes and parsley. Cover; simmer 4-5 hours. When ham is tender, remove meat from bone; cut in pieces; return to soup. Season to taste with salt and pepper.

Mrs. A. E., Minn.

CURRIED CHICKEN SOUP

¼ c butter, melted	⅛ tsp. pepper
½ c diced celery	1-1½ tsp. curry powder
¼ c minced onion	1 qt. milk
¼ c flour	2 c diced, cooked chicken
1 tsp. salt	¼ c chopped parsley

Saute celery and onion in butter over low heat until tender. Blend in flour and seasonings. Add milk; cook, stirring constantly, until smooth and thickened. Add chicken; heat thoroughly. Add parsley just before serving.

Mrs. N. C., Billings Co., N. D.

EASY VEGETABLE BEEF SOUP

¼ c chopped green pepper
½ c chopped onion
2 tbsps. fat
½ lb. ground beef
1 c tomato juice
½ c diced potatoes
½ c diced carrots
1½ tsps. salt
2 tbsps. butter
2 tbsps. flour
2 c milk

Cook green pepper and onion juice in fat until tender. Add ground beef; cook until browned, breaking up thoroughly. Drain off excess fat; add tomato juice, potatoes, carrots and salt. Cover; simmer for 20 minutes or until vegetables are tender. Melt butter; add flour and blend. Add milk gradually. Cook, stirring constantly, until thickened. Blend with soup mixture. Heat and serve.

Mrs. H. H., Neb.

OLD-FASHIONED BEAN SOUP

2 c navy beans
water
2 c cubed ham
ham bone
6 c water
1½ tsps. salt
¼ tsp. pepper
½ tsp. parsley flakes
1 c diced celery
1 large onion
1½ c tomato juice

Soak beans overnight in water. In morning add cubed ham, ham bone, 6 cups water, salt and pepper. Cover and simmer 2 hours. Add remaining ingredients; simmer 1 hour.

Mrs. B. A., Golden Valley, N. D.

CORN SOUP

2 slices onion
2 c cut, cooked yellow corn
1 qt. milk, scalded
3 tbsps. butter
2 tbsps. flour
½ tsp. salt
1/16 tsp. pepper
1/16 tsp. nutmeg
bacon

Combine onion, corn and scalded milk. Blend remaining ingredients, except bacon; add to milk mixture. Stir over low heat until slightly thickened. Discard onion. Serve with crumbled crisp bacon.

Mrs. B. N., Barnes Co., N. D.

COUNTRY CHOWDER WITH PORK

½ lb. dried lima beans
2 small, fresh pork hocks
1 tbsp. salt
¼ tsp. pepper
1 bay leaf
1 large potato, diced
2 large carrots, diced
1 large onion, diced
1 c cabbage, cut in small pieces
½ c tomatoes
½ c cut green beans

Soak Lima beans in 4 quarts cold water for 2 hours. Add pork hocks, salt, pepper and bay leaf. Simmer slowly for 1½ hours. Add remaining ingredients. Cook 30 minutes over low heat; serve.

Mrs. A. K., Michigan

VEGETABLE CHOWDER

3 tbsps. butter, melted
¼ c chopped onion
1 c sliced celery
2 tbsps. flour
1 tsp. salt
⅛ tsp. pepper
2½ c milk
1 (10-oz.) pkg. frozen Lima beans, cooked
1 (1-lb.) can tomatoes, undrained
1 (1-lb.) can cream style corn

Saute onion and celery in butter until onion is soft, but now brown. Blend in flour, salt and pepper. Add milk; cook, stirring constantly,until thickened. Add beans, tomatoes and corn; heat.

Mrs. M. S., Dewey Co., S. D.

SANDWICHES

IRISH BEEF BURGERS

1 (12-oz.) can corned beef, flaked
3 tsps. finely chopped onion
1½ tbsps. prepared mustard
3 tbsps. salad dressing
2 tsps. horseradish
8 sandwich buns
8 slices pasteurized process American cheese

Combine all ingredients except buns and cheese. Spread mixture on bottom halves of buns; top each with one slice cheese. Broil under low heat or bake at 350 degrees for 5 minutes to melt cheese. Top with other bun half and serve.

Mrs. P. R., Beadle Co., S. D.

SANDWICHES

HOT GROUND MEAT SANDWICHES

1 lb. bolgna, ground
¼ lb. cheddar cheese, ground
chopped pickle
chopped onion
mayonnaise

Mix together all ingredients. Spread on bun. Wrap in tin foil. Bake at 350 degrees to heat thoroughly and melt cheese. Serve hot.

V. K., Brown Co., S. D.

LUAU BURGERS

Shape ground beef into patties. Place each patty on square of aluminum foil. Top with a pineapple ring; fill the hole with chili sauce or barbecue sauce. Wrap in foil. Broil in oven or over charcoal until done. Serve on bun.

J. K., Mercer Co., S. D.

PIZZA BURGERS

1 (12-oz.) can Spam
½ lb. sliced Mozzarella cheese
1 medium onion
1 (15-oz.) can pizza sauce
buttered buns

Chop Spam, Mozzarella cheese and onion in food chopper. Add pizza sauce. Spread on buttered buns; broil until bubbly. Cover with slightly browned bun top.

Mrs. V. F., Stutsman Co., N. D.

LUMBERJACK HAMBURGERS

1 3 c butter or margarine
2 c finely chopped onions
1 c finely chopped green pepper
3 lbs. ground beef
2 eggs, slightly beaten
3 tbsps. Worcestershire sauce
2 tsps. salt
½ tsp. pepper
¾ c chili sauce
1½ tsps. oregano
1 c grated sharp Cheddar cheese

Melt butter in skillet. Add onions and green pepper; cook until tender, about 5 minutes. Combine ground beef, eggs, Worcestershire sauce, salt and pepper. Mix lightly. Shape into 12 long thin patties, about 6x3 inches. Spoon green pepper mixture onto centers of six patties. Top each with 2 tablespoons of chili sauce and ¼ teaspoon oregano. Cover with remaining patties. Press and seal edges together securely. Broil 3 inches from heat for about 5 minutes on each side, or until meat is done. Sprinkle cheese over top of patties and broil until cheese melts.

Mrs. A. J. K., Meade Co., S. D.

HOT MEAT SANDWICHES

1 lb. minced ham, ground
¾ lb. cheese, ground
¼ c mustard
1/3 c salad dressing
2 tsps. sweet pickles
1 tsp. onion
2 hard boiled eggs, chopped (optional)
hamburger buns

Combine all ingredients except hamburger buns; mix well. Spread mixture on hamburger buns; wrap in aluminum foil Bake at 325 degrees for 25 minutes. Serve hot.

Mrs. S. P., Dawson Co., Mont.

TUNA BURGERS

1 can cream of mushroom soup
1 (7-oz.) can tuna, drained and flaked
1 tbsp. minced green pepper
1 tbsp. minced onion
6 hamburger buns

Combine first four ingredients; spread mixture onto hamburger buns. Brush top half of bun with butter. Bake on cookie sheet at 400 degrees for about 10 minutes. Serve at once.

Mrs. R. T., Jr., Lake Co., S. D.

BUN FILLING

1 lb. luncheon loaf or ground ham
½ lb. cheese, ground or grated
1 c tomato paste
1 c softened butter
1 small onion (optional)

Mix together all ingredients; spread on unbuttered buns; wrap each bun in foil and heat.

Mrs. E. T., Roberts Co., S. D.

PICNIC CHEESE SANDWICH

1 tbsp. butter, melted
1 tbsp. flour
1 tbsp. sugar
1 egg, well-beaten
½ c light cream
2 hard boiled eggs, chopped
¼ lb. cheese, diced
1 tbsp. chopped pimiento
¾ tsp. salt
⅛ tsp. pepper
whole wheat bread
white bread

Combine butter, flour, sugar, egg and cream. Cook over low heat until boiling; stir frequently. Combine chopped eggs, cheese, pimiento, salt and pepper. Pour hot sauce over egg mixture; mix thoroughly. Store in refrigerator until ready to use. Spread on slices of buttered whole wheat bread; cover with white bread.

Mrs. S. B., Campbell Co., S. D.

SANDWICHES

FRENCH-TOASTED CHICKEN SANDWICHES

1 c finely chopped, cooked or canned chicken
½ c minced celery
¼ c chopped sweet pickle
1 tsp. salt
pepper
¼ c mayonnaise
12 thin slices white bread
3 eggs, beaten
½ c milk
1 tsp. sugar
¼ tsp. salt
butter or margarine

Combine chicken, celery, pickle, seasonings and mayonnaise; mix well. Make up 6 sandwiches with bread and set aside. Mix eggs, milk, sugar and salt in pie pan. Dip each sandwich into egg mixture, being sure to coat each side completely. Saute sandwiches slowly in butter in hot skillet on both sides until filling is thoroughly heated and sandwiches are golden brown.

Mrs. P. C., Carter Co., Mont.

CRUNCHY HAM-CHEESE SANDWICHES

4 c oven-toasted rice cereal
4 thin slices cooked ham
4 slices processed Swiss cheese
8 slices day-old bread
3 eggs, slightly beaten
2/3 c milk
¼ tsp. salt
¼ c butter or margarine, melted

Crush cereal to 2 cups; set aside. Place a slice of ham and a slice of cheese between 2 slices of bread. Cut sandwiches in half diagonally. Set aside. Combine eggs, milk and salt in shallow dish or pan. Dip sandwiches in egg mixture, turning once. Coat evenly with slightly crushed cereal. Place flat side down in single layer on well-greased baking sheet. Drizzle with melted margarine. Bake at 450 degrees for 10 minutes or until crisp and lightly browned. Serve hot.

Ms. C. W., Slope Co., N. D.

SALADS & DRESSINGS

What's in your refrigerator right now? Probably the makings of a good salad. Some of the tastiest salads ever created have been the product of some inventive housewife's imagination. From grandma's fresh garden salads, to the intricate dishes of today, there's plenty of nutritious and delicious eating in the makings of a good salad. Try some of the ones included here . . . they're home-tested and delicious.

Vegetable
Fruit
Salad Dressings
Main Dish
Relishes

VEGETABLE COMBINATIONS

1. Cabbage, dates and nuts with sour cream dressing.
2. Shredded or chopped cabbage, peas and cheese with favorite dressing.
3. String beans and sweet onions.
4. Cooked kidney beans, celery and chopped pickle.
5. Shredded cabbage, grated raw carrots, raisins and mayonnaise.
6. Cabbage, green pepper, celery and raw carrots.
7. Cabbage, horseradish and beets.
8. Cold boiled lima beans, parsley, pimientos and onion.
9. Asparagus tips and green or red pepper rings.

Mrs. E. S., Mercer Co., N. D.

COLE SLAW

2 qts. shredded cabbage
sweet onion, sliced thin
1 c cream, whipped
3 tbsps. sugar
3 tbsps. vinegar
½ c grated carrots

Mix together cabbage and sweet onion. Chill ½ hour. Add sugar and vinegar to whipped cream. Fold into cabbage mixture. Sprinkle carrots over top.

Mrs. L. H., Davison Co., S. D.

CUCUMBER DELIGHT SALAD

1 large cucumber
1 small onion, chopped
1 c creamed cottage cheese
¼ c salad dressing
¼ c dairy sour cream
1 pkg. unflavored gelatin
¼ c cold water
1 pkg. lime gelatin
¾ c boiling water
4 ice cubes

Peel cucumber; shred, drain and discard juice. Add onion to cottage cheese, salad dressing and sour cream. Add cucumber and stir to mix. Dissolve unflavored gelatin in cold water. Add lime gelatin and boiling water; stir to dissolve. (You may have to place on hot burner to thoroughly dissolve.) Add ice cubes; stir until melted. Add cucumber mixture; stir to mix well. Chill until firm.

Mrs. A. A. B., Hutchinson Co., N. D.

THREE-BEAN SALAD

1 can cut green beans
1 can cut wax beans
1 can kidney beans
½ c chopped green pepper
½ c chopped onion
¾ c sugar
2/3 c vinegar
1/3 c salad oil
1 tsp. salt
1 tsp. pepper

Drain and mix together beans; add green pepper and onion. Mix together remaining ingredients. Pour over bean mixture; marinate overnight. Drain and serve.

Mrs. A. V., Douglas Co., S. D.

SALADS/Vegetable

SAUERKRAUT SALAD

1 or 2 cans sauerkraut
½ c sugar
½ c shredded carrots
½ c chopped celery
2 tbsps. chopped pepper (green or red)
¼ c chopped onion

Cut sauerkraut into pieces. Add sugar; let stand ½ hour. Add remaining ingredients. Cover and refrigerate several hours.

Variation: Add 1 can pimiento or caraway seeds.

Mrs. C. S. B., Beadle Co., S. D.

OLD-FASHIONED POTATO SALAD

6 c cool, diced boiled potatoes
¼ c finely chopped onion
1 tsp. salt
¼ tsp. pepper
1 tsp. celery seed
¼ c vinegar
¼ c sugar
9 hard boiled eggs, cut up
salad dressing

Combine potatoes, onion, salt, pepper, celery seed, vinegar and sugar and mix lightly. Add eggs. Add enough salad dressing to moisten.

Mrs. T. Y., Ward Co., N. D.

JELLIED CUCUMBER-CABBAGE SALAD

1 pkg. lime gelatin
1 c hot water
½ c mayonnaise
½ tsp. celery seed
1 c finely chopped cabbage
½ c chopped cucumber
1 small onion, diced
2 tbsps. chopped green pepper

Dissolve gelatin in hot water; let stand to cool. When cool, whip in mayonnaise; add chopped vegetables and celery seed. Pour into oblong pan, dish or individual molds and refrigerate until firm.

Mrs. E. S., Mercer Co., N. D.

SALADS/Fruit

APPLE SAUCE SALAD

1 c hot water
1/3 c red cinnamon candy
1 pkg. lemon gelatin
1½ c sweetened apple sauce
1 (8-oz.) pkg. cream cheese
½ c diced celery
½ c chopped nut meats
½ c salad dressing

Combine hot water and cinnamon candy. Let candy dissolve; add lemon gelatin. Stir to dissolve. Add apple sauce. Pour half this mixture in a 12x9x2-inch pan. Chill until firm. Blend together remaining ingredients; pour over firm apple mixture. Pour remaining apple mixture over cream cheese mixture; chill until firm.

Mrs. J. H., Hettinger Co., N. D.

CRANBERRY ORANGE SALAD

- 2 pkgs. orange gelatin
- 1 c boiling water
- 2 cans (1 lb. each) jellied cranberry sauce, mashed
- ¾ c orange juice
- 2 tsps. grated orange rind (optional)
- 1 c chopped peeled apples
- ½ c chopped celery
- salad greens
- 1 pt. creamed cottage cheese
- ¼ c slivered toasted almonds (optional)

Dissolve gelatin in boiling water. Add cranberry sauce and orange juice; mix well and cool until consistency of egg white. Stir in grated orange rind, chopped apple and celery. Pour into 1¼-quart (5-cup) ring mold. Chill until firm. Unmold on crisp salad greens; fill center of ring with cottage cheese and top with almonds if desired.

Ms. D. O., Campbell Co., S. D.

FROSTED CRANBERRY SALAD

- 1 (13½-oz.) can crushed pineapple
- 2 (3-oz.) pkgs. lemon gelatin
- 1 (7-oz.) bottle ginger ale
- 1 (1-lb.) can jellied cranberry sauce
- 1 (2-oz.) pkg. dessert topping mix
- 1 (8-oz.) pkg. cream cheese, softened
- ½ c chopped pecans

Drain pineapple, reserving syrup. Add water to make 1 cup; heat to boiling. Dissolve gelatin in hot liquid; cool. Gently stir in ginger ale; chill until partially set. Blend fruits; fold into gelatin. Turn in 9x9x2-inch pan; chill until firm. Prepare topping mix according to package directions. Blend in cheese; spread over gelatin. Toast nuts in 1 tablespoon butter at 350 degrees for 10 minutes. Sprinkle on top. Chill.

Mrs. V. F., Faulk Co., S. D.

ORANGE SHERBET SALAD

- 2 pkgs. orange gelatin
- 1 c hot water
- 2 c orange sherbet
- 1 can mandarin oranges, drained
- 2 c whipped cream

Dissolve gelatin in hot water; add sherbet and stir until dissolved. Fold in mandarin oranges and whipped cream; place in mold and refrigerate.

Mrs. R. S., Minnehaha Co., S. D.

5-CUP SALAD

- 1 c miniature marshmallows
- 1 c crushed pineapple, drained
- 1 c mandarin orange sections, drained
- 1 c flaked coconut
- 1 c cultured sour cream
- maraschino cherries

Mix together all ingredients except maraschino cherries. Dot with cherries to give color.

Mrs. A. H., Brookings Co., S. D.

PINEAPPLE NOEL SALAD

2 envelopes unflavored gelatin
2 tbsps. sugar
¼ tsp. salt
3½ c canned pineapple juice
2 tbsps. lemon juice
1 (11-oz.) can mandarin oranges, drained
3 bananas
½ c thinly sliced celery
pineapple slices
salad greens

Mix gelatin, sugar and salt. Add ½ c pineapple juice and place over low heat, stirring constantly, until gelatin is dissolved. Stir in remaining pineapple juice and lemon juice. Cool until mixture thickens slightly. Add oranges, sliced bananas and celery into thickened gelatin. Turn into 5-cup mold. Chill until firm. Unmold. Garnish with pineapple slices and salad greens.

Mrs. L. H., Charles Mix Co., S. D.

FRUIT SALAD

1 can peach pie filling
2 c miniature marshmallows
1 c chunk pineapple, drained
1 c seedless grapes
1 red apple, diced
red maraschino cherries
bananas

Combine peach pie filling and marshmallows. Let stand about ½ hour. Add pineapple, grapes, apple and cherries. Mix well and chill 3-4 hours. Slice in a few bananas before serving.

Mrs. W. K., Day Co., S. D.

STRAWBERRY SALAD

2 pkgs. strawberry gelatin
2 c boiling water
2 pkgs. frozen strawberries
1 medium can crushed pineapple
2 large bananas, mashed
1 c whipped cream

Dissolve gelatin in boiling water; add undrained strawberries, pineapple and mashed bananas. Pour half this mixture into an 8x12-inch pan. Let set. Spread whipped cream over set gelatin; pour remaining gelatin mixture on top.

Mrs. E. K., McCook Co., S. D.

GLORIFIED RICE

1½ c boiled rice, rinsed in cold water
1 c crushed pineapple drained
1 c fruit cocktail, drained
24 marshmallows, cut-up
¼ c sugar
1 c whipped cream

Mix together all ingredients except whipped cream. Let stand 1 hour. Fold in whipped cream and serve.

Mrs. O. B., Jerauld Co., S. D.

FROZEN FRUIT SALAD

- 1 (3-oz.) pkg. cream cheese
- 3 tbsps. cream
- 1/3 c mayonnaise
- 2 tbsps. lemon juice
- 1 c pineapple tidbits, drained
- 1 c mandarin orange sections, drained
- ½ c cherries, pitted and halved
- ½ c maraschino cherries, quartered
- ½ c chopped pecan meats
- 2 tbsps. sugar
- 1 c whipping cream, whipped

Mix softened cream cheese with cream. Add mayonnaise and lemon juice; mix well. Add remaining ingredients except whipped cream. Fold in whipped cream. Ladle into a quart milk carton and allow to freeze without stirring. About 30 minutes before serving, remove from freezer and put in refrigerator. Cut open carton and slice salad.

Note: Add red food coloring before folding in whipped cream if desired.

Miss E. K., McCook Co., S. D.

Main Dish/SALADS

CHICKEN SALAD

- 3 c diced cold chicken
- 1 c cooked, cooled shell macaroni
- 1 c peas
- 1 c diced celery
- 5 hard boiled eggs
- 1 tbsp. pimiento

Toss together all ingredients lightly. Moisten with a mayonnaise dressing.

Mrs. H. D., Brown Co., S. D.

CHEESE AND MACARONI SALAD

- 2 c cooked macaroni
- 1 c chopped celery
- 1 green pepper, chopped
- 6 sweet pickles, chopped
- 2 pimientos, chopped
- 1 c grated cheese
- mayonnaise or French dressing

Blend all ingredients well. Chill before serving.

Mrs. F. K. M., Ward Co., N. D.

TUNA OR CHICKEN SALAD

- 3 c cooked rice
- 1 c mayonnaise
- 1 tsp. salt
- 1½ c peas
- 1½ c tuna or diced chicken
- 1½ c diced celery
- stuffed olives
- 3 hard boiled eggs

Mix together all ingredients. Chill 2 hours; serve.

Mrs. J. H., Hettinger Co., N. D.

SALADS/Main Dish

TUNA SALAD

1½ c cooked macaroni, drained and cooled
1 c peas, drained
2 tbsps. chopped pimiento
1 can tuna
1 small onion, diced
½ tsp. pepper
1 c mayonnaise
cream
diced celery (optional)

Mix together macaroni, peas, pimiento, tuna, onion and pepper. Dilute mayonnaise with small amount of cream and add to tuna mixture. Diced celery may be added if desired.

Mrs. J. S., Ward Co., N. D.

MACARONI SALAD

1 pkg. cooked ring macaroni, cooled
1 can peas
1 c chopped celery
1 small onion, finely chopped
1 c chicken, ham, tuna or salmon
3 hard boiled eggs, finely chopped
mayonnaise
salt
pepper

Combine all ingredients; season to taste.

Mrs. G. B., Marshall Co., Minn.

SALADS/Relishes

CANDIED DILL PICKLES

3 c sugar
2/3 c cider vinegar
2 tbsps. mixed pickling spices
1 qt. dill pickles

Combine sugar, vinegar and pickling spices; bring to full boil for 1 minute. Cool slightly. Drain dill pickles; cut into ¼-inch slices and repack in jar. Pour warm syrup over pickles. Refrigerate for 1 week before use.

Mrs. H. A., Hamlin Co., S. D.

RAW CARROT RELISH

2 c grated carrots
1 tbsp. sugar
pinch of salt
juice of 1 lemon
2 oranges

Mix together ingredients well. Pack in glass jar and refrigerate several hours. Serve with potatoes. This keeps well for several days.

Mrs. J. K., McPherson Co., S. D.

APPLE RINGS

1 c sugar
¾ c water
¼ c vinegar
5 whole cloves
1 (1-inch) stick cinnamon
apples
food coloring

Combine sugar, water, vinegar, cloves and cinnamon; cook for 5 minutes. Wash apples; do not peel. Core and cut crosswise into ½-inch slices. Cook slowly in syrup until transparent. Remove carefully. Add food coloring to syrup and cook until thick. Pour over apple slices. Cool. Remove from syrup and use as garnish.

Mrs. G. R.. Stark Co., N. D.

RELISH

2 medium-sized heads of cabbage
8 carrots
8 green or red peppers
12 onions
½ c salt
6 c sugar
6 c vinegar
½ tsp. celery seed
½ tsp. mustard seed

Grind together cabbage, carrots, peppers and onions. Add salt and let stand 2 hours. Drain and add remaining ingredients. Don't cook. Place in small jars and keep in cool place for storage. Does not have to be sealed.

Mrs. V. W. S., Custer Co., S. D.

Dressings/SALADS

TART TOSSED SALAD DRESSING

½ c sugar
1 tsp. dry mustard
1 tsp. salt
1 tbsp. grated onion
1/3 c vinegar
1 c salad oil
1 tsp. celery seed
¼ tsp. paprika

Mix together all the ingredients; beat well. This will thicken and keep well in the refrigerator. This is a good dressing for fresh cabbage or a tosssed lettuce salad.

Mrs. E. S., Mercer Co., N. D.

HONEY APPLE DRESSING

½ c apple juice
¼ c honey
1 tbsp. cornstarch
pinch of salt
2 beaten egg yolks
½ c whipping cream, whipped

Blend together apple juice, honey, cornstarch and salt in top of double boiler. Cook until mixture begins to thicken. Stir small amount of the hot mixture into egg yolks. Combine mixtures and cook over low heat until thick (about 2 minutes). Chill thoroughly. Fold cold honey mixture into whipped cream. Serve with fruit salads.

Ms. F. N., Brookings Co., S. D.

PINEAPPLE FRUIT SALAD DRESSING

- 1 tsp. butter
- ½ c pineapple juice or pineapple-orange juice
- 1 tbsp. lemon juice
- ¼ c sugar
- 2 egg yolks, beaten
- ½ c whipping cream, whipped

Melt butter in top of double boiler. Add fruit juice, egg yolks and sugar; cook until thick, stirring constantly. Cool. Fold whipped cream into cooled mixture just before serving.

Mrs. W. A., Barnes Co., N.D.

SOUR CREAM DRESSING

- ½ c sour cream
- 2 tbsps. sugar
- ½ tsp. salt
- 3 tbsps. vinegar or lemon juice
- dash of paprika
- dash of garlic salt
- dash of onion salt
- dash of celery salt

Beat the cream until stiff. Add remaining ingredients slowly and continue beating until thick. Good with lettuce or cabbage.

Mrs. E. S., Mercer Co., N. D.

THOUSAND ISLAND DRESSING

- 2 c salad dressing
- 1 c tartar sauce
- 1 can tomato soup
- 1 small onion, grated
- 1 clove garlic (optional)
- ¼ c chopped sweet pickle
- 1 tsp. dry mustard
- 2 tsps. prepared mustard
- 1 tbsp. sugar
- 2 tbsps. chili sauce
- 2 tbsps. vinegar
- 2 tbsps. sweet pickle juice

Mix together all ingredients; let stand 24 hours before serving. Store in refrigerator.

Mrs. M. R., Spink Co., S. D.

TOMATO DELIGHT SALAD DRESSING

- 1 can tomato soup
- ¾ c salad oil
- ¾ c vinegar (plain or wine)
- ½ c sugar
- 2 tbsps. Worcestershire sauce
- 1 tbsp. salt
- 1 tsp. black pepper
- 2 cloves garlic

Mix together all ingredients. This recipe needs no refrigeration and will not spoil. Leave the garlic in as long as dressing lasts.

Mrs. K. L., Spink Co., S. D.

VEGETABLES

What farm is complete without a large garden to provide fresh foods for summer eating and line pantry shelves with jars of vegetables for the winter months? Fresh from the garden green beans served with cheese sauce, hot sweet corn dripping with butter, new potatoes and peas . . . vegetables supply the colorful accompaniment that rounds out hearty meals. Give the farm-tested recipes in this section a try; many of them are sure to become family favorites.

Asparagus
Beans, Baked
Beans, Green
Beans, Lima
Beets
Cabbage
Carrots
Onions
Peas
Peppers
Potatoes
Rice
Squash
Sweet Potatoes
Turnips
Zucchini
Mixed Vegetables
Sauces

VEGETABLE TIME TABLE

Vegetable	*Preparation*	*Time*	*Pressure*
Artichokes, French	Cut off 1 inch of top, stem and tips of leaves. Cook upright in boiling salted water to cover.	20-30 min.	10 min.
Jerusalem	Wash, pare; leave whole or slice.		1-2 min. (cut-up)
Asparagus	Snap off woody ends, remove scales if sandy. Break stalks. Cook cut-up or whole in small amount of boiling water.	10-15 min. (whole) 8-10 min. (cut-up)	
Beans, Green or waxed	Wash; remove ends and strings. Cook whole or cut-up in small amount of boiling salted water.	20-30 min.	3-4 min.
Limas	Shell and wash. Cook covered in small amount of boiling salted water.	20-30 min.	2-3 min.
Beets	Cut off all but 1 inch of stems and root. Cook covered in boiling salted water; peel when done.	35-60 min.	10-18 min.
Broccoli	Remove outer leaves and tough parts of stalks. Split rest of stalk almost to flowerets. Or cut in 1-inch pieces.	15-20 min. (split) 10-15 min. (cut-up)	2-3 min.

Vegetable	*Preparation*	*Time*	*Pressure Cooker*
Brussels Sprouts	Cut off wilted leaves. Wash. Cut large sprouts in half lengthwise. Cook covered in small amount of boiling salted water.	10-15 min.	3 min.
Cabbage, Green	Cut off wilted outer leaves. Cut into 6-8 wedges, or shred.	10-12 min. (wedges) 5-7 min. (shredded)	3-4 min.
Carrots	Wash and pare. Leave whole, slice or cut in quarters. Cook covered in small amount of boiling salted water.	20-25 min. (while) 15-20 min. (cut-up)	4-8 min. (whole) 3 min. (sliced)
Cauliflower	Remove leaves and part of woody stem. Leave whole or separate into flowerets. Cook covered in small amount of boiling salted water.	20-25 min. (whole) 10-15 min. (cut-up)	5 min. (whole) 2 min. (cut-up)
Chard, Swiss	Wash. Cook covered in small amount of boiling salted water.	10-20 min.	2 min.
Corn On Cob	Remove husks and silks. Rinse. Cook covered in small amount of boiling salted water.	6-8 min.	3-5 min.
Eggplant	Wash; pare if skin is tough. Cut into ½-inch slices. Dip in beaten egg; dip in dry bread crumbs. Brown slowly on both sides in small amount of hot fat.	4 min.	
Leeks	Cut off green tops to 2 inches from white part. Wash; cook covered in small amount of boiling salted water.	15-20 min.	
Lentils, dried	Wash. Add 2½ times as much water as lentils. Simmer covered.	35 min.	
Onions	Peel under water. Slice, cut in quarters, or leave small onions whole. Cook covered in small amount of boiling salted water.	25-35 min.	5-7 min. (whole)

Parsnips	Wash; pare; slice crosswise or lengthwise. Cook covered in small amount of boiling salted water.	15-20 min.	7 min.
Peas			
Green	Shell and wash. Cook covered in small amount of boiling salted water.	8-15 min.	1 min.
Black-eyed	Rinse; add 2½ times as much water as peas. Soak overnight. Add salt. Cook in water used for soaking.	35-45 min.	
Potatoes	Wash. Cook with skins or wash and pare. Cook whole, quarter, or cube. Whole — cook covered in boiling salted water to cover. Cut-up — Cook covered in small amount of boiling salted water.	25-40 min. (whole) 20-25 min. (quartered) 10-15 min. (cubed)	10 min.
Rutabagas	Wash; pare. Slice or cube. Cook covered in small amount of boiling salted water. Mash if desired.	25-40 min.	3-5 min.
Spinach	Cut off roots; wash in lukewarm water. Cook in just water that clings to leaves. Reduce heat when steam forms.	3-5 min.	1-3 min.
Squash, Acorn	Wash. Cut in half. Bake cut side down at 350 degrees 35-40 min.; turn cut side up. Bake until done. Or pare and cube. Cook covered in small amount boiling salted water.	50-60 min. total 15 min. (cubed)	
Hubbard	Wash; cut in serving size pieces. Do not pare. Wrap in foil with butter. Bake at 350 degrees. Or pare and cube. Cook covered in small amount boiling salted water.	1¼ hrs. (baked) 15 min. (boiled)	10-12 min.
Summer	Wash; pare. Slice or cube. Cook covered in small amount boiling salted water.	15-20 min.	5-7 min.
Zucchini	Wash; do not pare. Slice thin. Cook covered in butter in skillet 5 minutes; uncover and cook until tender.	10 min. total	5-7 min.
Tomatoes	Wash. Plunge in boiling water; cool under cold water. Peel and cut up. Cook slowly, covered; do not add water. Season.	10-15 min.	1 min.
Turnips	Wash; pare. Slice or cube. Cook covered in small amount boiling salted water. Mash, if desired.	15-20 min.	3-5

VEGETABLES/Asparagus

ASPARAGUS IN CASSEROLE

3 c fresh asparagus
3 tbsps. butter
3 tbsps. flour
1 c asparagus liquid or milk
½ tsp. salt
½ tsp. grated pasteurized process cheese
2 c large bread crumbs
melted butter

Cook asparagus until tender, adding ½ teaspoon salt toward the end of cooking time; drain well. Cut into 1-inch pieces; place in greased pie pan. Melt butter; add flour, then liquid a little at a time, stirring constantly. Cook until thick; add salt and grated cheese. Cook until cheese is melted. If sauce is too thick add more liquid. Pour sauce over asparagus. Cover top with bread crumbs and drizzle with melted butter. Bake at 325 degrees for 30 minutes. To brown crumb topping place under broiler for several minutes.

Mrs. A. L., Golden Valley Co., N. D.

CREAMY ASPARAGUS

¼ c butter
¼ c flour
2 c milk
¼ tsp. salt
1 (3-oz.) pkg. cream cheese, crumbled
1 lb. fresh asparagus or 1 (10-oz.) pkg. frozen asparagus, cooked
4 slices toast
4 slices crispy cooked bacon, crumbled

Melt butter over low heat; blend in flour. Slowly add milk and cook until thick, stirring constantly. Add salt and cream cheese; stir until blended. Arrange hot asparagus on toast slices and pour cheese sauce over all. Garnish with crumbled bacon.

Mrs. R. T., Minnehaha Co., S. D.

VEGETABLES/Beans

BAKED BEANS

2 c navy beans
¼ lb. salt pork or bacon
1½ tsps. salt
¼ c brown sugar
2 tbsps. molasses
½ tsp. dry mustard
1 small onion, chopped
1 c tomato puree or 1 c tomato soup

Wash beans and cover with 2 quarts water. Soak overnight. Cook slowly until skins burst, drain and save liquid. Arrange beans and sliced bacon or salt pork alternately in casserole. Add all other ingredients to reserved liquid and pour over beans. Bake at 275 degrees for 6 hours.

E. H., Griggs Co., N. D.

STRING BEANS WITH CASHEW NUT BUTTER

- 3 (9-oz.) pkgs. frozen green beans or 5 c fresh green beans
- ½ c chopped cashews
- ½ c butter

Cook beans. Saute nuts in butter until light golden brown. Toss nuts with drained beans. Serve hot.

Mrs. J. S., McLean Co., N. D.

GREEN BEAN CASSEROLE

- 1 can green beans, drained
- 1 can cream of mushroom soup
- 1 can French fried onions

Place beans in casserole; add mushroom soup. Bake at 350 degrees for 20 minutes. Place onion rings on top. Bake at 400 degrees for 4 minutes.

Mrs. J. L., Brown Co., S. D.

GREEN BEAN BAKE

- 1 can cream of mushroom soup
- ½ c milk or light cream
- 1 tbsp. butter
- 1 tbsp. flour
- ½ tsp. Worcestershire sauce (optional)
- ½ c grated sharp cheese
- 1 qt. canned green beans
- 1 tbsp. chopped pimento
- ½ c buttered bread crumbs

Combine soup and milk or cream; heat. Melt butter and blend in flour. Add to soup mixture. Add Worcestershire sauce. Cook until thick, stirring constantly. Add cheese; stir until melted. Combine with beans and pimento. Place in casserole and sprinkle with bread crumbs. Bake at 350 degrees for 20 minutes.

Mrs. J. L., Morton Co., N. D.

CREAMED BEANS WITH ONIONS

- 3 tbsps. thinly sliced onion
- 1 tbsp. shortening
- 1 c cream
- 1½ tbsps. flour
- cooked green beans

Brown onion in shortening. Add cream and flour. Add cooked beans and heat until thickened.

Mrs. J. S., McLean Co., N. D.

VEGETABLES/Beans

LIMA BEAN POT

2 c dried Lima beans
1 lb. link pork sausage cut in ½-inch pieces
1 medium onion, chopped
½ medium green pepper, diced
½ c chopped celery
1¼ c tomato juice
1½ tsps. salt
1 tsp. chili powder
2 tbsps. brown sugar
1½ tsps. cayenne pepper (optional)

Soak Lima beans according to package directions. Cook until tender and most of cooking water has been absorbed. Brown sausage pieces lightly, drain off grease, reserving 2 tablespoons. Cook onion, green pepper and celery in reserved grease until onion is light golden brown and green pepper and celery are tender. Add to Lima beans, together with tomato juice, brown sugar and seasonings. Mix thoroughly. Cook over low heat for 1 hour.

Mrs. K. G., Pennington Co., S. D.

VEGETABLES/Beets

SPICED BEETS

8-10 small beets
½ c vinegar
1 c sugar
1 c water
1 tsp. cloves
1 tsp. allspice
1 tbsp. cinnamon

Boil beets until tender; cool and slip off skins. Combine remaining ingredients; add beets. Cover and simmer 15 minutes. Serve hot or cold.

Mrs. R. D., Foster Co., N. D.

VEGETABLES/Cabbage

CABBAGE WITH SOUR-CREAM SAUCE

1 medium head cabbage
boiling water
½ tsp. salt
1 c sour cream
¼ c cabbage liquid
½ tsp. salt
1 tbsp. fresh chives

Wash cabbage and cut into wedges. Place in saucepan with 1-inch boiling water and ½ teaspoon salt. Cook, uncovered, 5 minutes. Cover and cook until crisp-tender, about 8-10 minutes. Place wedges on warm serving platter. Heat sour cream with ¼ cup of the cabbage liquid and salt, but do not boil. Pour over cabbage. Garnish with chives.

Mrs. D. L., Davison Co., S. D.

GLAZED CARROTS AND PINEAPPLE

½ c pineapple chunks
½ c water
2½ c cut-up carrots
¾ tsp. salt
1 tbsp. cornstarch
2 tbsps. butter or margarine

Drain pineapple; reserve syrup. Bring water to a boil; add carrots and salt; cover and bring to a boil again. Reduce heat and simmer until tender. Drain any remaining liquid into a measuring cup; add pineapple syrup to make 1 cup. Mix cornstarch and liquid in a saucepan; cook, stirring constantly, until thick and clear. Stir in butter. Add carrots and pineapple; heat to serving temperature. Serve hot.

M. J. P., Lake Co., S. D.

SCALLOPED CARROTS

10 carrots
1 c bread crumbs
½ c chopped onion
½ c cream
½ c milk
salt and pepper
3 tbsps. butter
3 eggs, beaten

Peel and slice carrots; boil until soft. Drain and mash. Combine carrots and remaining ingredients. Place in casserole. Bake at 350 degrees for 30 minutes.

Mrs. K. B., Burleigh Co., N. D.

Onions/VEGETABLES

CREAMED ONIONS WITH CHESTNUTS

1½ lbs. small "boiler" white onions
1 c medium cream sauce
½ lb. boiled chestnuts

Wash, peel and cook onions uncovered in large amount of boiling salted water for 20 minutes, or until tender. Drain and dress with cream sauce. Add boiled chestnuts that have been peeled, scraped and cut into large pieces.

Mrs. E. M. B., Brookings Co., S. D.

Peas/VEGETABLES

BAKED PEAS

1 can peas, drained
6 slices of bacon
1 c cream
1 tsp. salt
⅛ tsp. pepper
½ c bread crumbs

Cut bacon in small pieces and brown. Drain excess fat. Add bacon to peas. Add cream and seasoning. Place in casserole and cover with bread crumbs. Bake 40 minutes.

Mrs. C. G., Stutsman Co., N. D.

VEGETABLES/Peas

PEAS CREOLE

¼ c butter or margarine
½ c chopped onions
½ c chopped green pepper
½ c sliced mushrooms
1 (16-oz.) can peas, drained
2 (8-oz.) cans tomato sauce
1 tsp. salt
⅛ tsp. pepper
cooked rice

Melt butter or margarine; add onions, green pepper and mushrooms. Cook until tender. Add peas, tomato sauce, salt and pepper. Cook 10 minutes, stirring occasionally. Serve over rice.

Mrs. J. K., Cavalier Co., N. D.

VEGETABLES/Green Pepper

GREEN PEPPERS STUFFED WITH FRESH VEGETABLES

6 medium green peppers
boiling water
3 medium carrots, finely chopped
1 large potato, finely chopped
3 stalks celery, finely chopped
2 c tomatoes, diced
2 tbsps. chopped parsley
2 tsps. salt
¼ tsp. ground black pepper
3 tbsps. butter or margarine
2 eggs, beaten
1 c fine, dry bread crumbs
6 tbsps. grated pasteurized process cheese

Cut tops off green peppers and scoop out seeds. Parboil 10 minutes in 1 inch boiling water in saucepan. Remove from water and drain. Mix carrots, potato, celery, tomatoes, parsley, salt and black pepper. Melt butter or margarine in skillet. Add chopped vegetables and cook until tender, about 15 minutes. Add beaten eggs and bread crumbs; spoon into green peppers. Bake at 375 degrees for 10-15 minutes, or until peppers are tender. Sprinkle tops of peppers with cheese and place under broiler 5 minutes or until cheese is melted and lightly browned.

Mrs. D. K., Adams Co., N. D.

potatoes

Potatoes are the backbone of nearly every meal. Seldom will you find a farm family in the Dakotas preparing to eat a noon or evening meal when there are no potatoes on the table.

North Dakota's Red River Valley and the area around Watertown, S. D., are famous for their fine quality potatoes. Dakota potato growers hope you will have potatoes with your next meal.

BAKED BAR-BE-Q POTATOES

1 tbsp. flour
1½ tsps. salt
⅛ tsp. pepper
4 c thinly sliced raw potatoes
½ c chopped onion
1 c shredded pasteurized process cheese
1/3 c ketchup
1 tsp. Worcestershire sauce
3 drops hot pepper sauce
1½ c scalded milk
2 tbsps. butter
4 slices bacon, diced, fried crisp and drained

Combine flour, salt and pepper; mix. Arrange potatoes, onion, flour mixture and half of the cheese in layers in shallow 2-quart casserole. Combine ketchup, Worcestershire sauce, hot pepper sauce and scalded milk; pour over potatoes and dot with butter. Cover; bake at 375 degrees for 45 minutes. Uncover and stir. Continue baking until potatoes are tender, 15-20 minutes. Sprinkle with remaining cheese and bacon. Serve immediately.

Ms. P. C. R., Lyon Co., Minn.

ITALIAN POTATOES

2 tbsps. bacon fat
1½ c boiled, cubed potatoes
1 tsp. minced onion
¼ tsp. minced parsley
1 egg yolk, beaten
⅛ tsp. paprika

Heat fat in frying pan. Add potatoes, onion and parsley; brown. Lower heat; add egg yolk and paprika. Cover and simmer for 5 minutes.

Miss A. L., Custer Co., Mont.

AU GRATIN POTATOES

2 tbsps. butter
2 tbsps. flour
1½ tsps. salt
2 c milk
6 medium boiling potatoes, pared and sliced
¼ lb. cheese, grated

Melt butter; add flour and salt; stir until smooth. Gradually add milk, cooking over direct heat and stirring constantly until mixture becomes thick and smooth. (If white sauce becomes lumpy beat with rotary beater.) Add potatoes to white sauce; heat gently until sauce bubbles. Arrange layers of creamed potatoes in buttered casserole with grated cheese between layers and on top. Cover and bake at 375 degrees for 25 minutes or until potatoes are tender. Remove cover and brown top, either in oven with temperature increased to 475 degrees or under broiler.

Mrs. A. R., Seward Co., Kan.

HOME-FROZEN FRENCH FRIES

Cut peeled potatoes lengthwise, then crosswise into 3/8-inch strips. Rinse in cold water, dry thoroughly between towels. Fry small amount in 360 degree deep fat about 5 minutes, until tender but not brown. Drain on paper toweling. Repeat with remaining batches. Cool to room temperature, then package and freeze.

To serve, potatoes may be browned at 550 degrees for 10 minutes or until brown, turning once; or fry in 390 degree deep fat about 1½ minutes. Season and serve immediately.

Mrs. E. W., McLean Co., N. D.

CHIP BEEF-POTATO SCALLOP

- ½ lb. dried beef
- 6 medium-sized potatoes
- ½ c flour
- 1 can cream of mushroom soup
- 2 c milk

Shred dried beef. Peel and slice potatoes. Alternate layers of potatoes, flour and beef in greased casserole, finishing with a layer of meat. Combine soup and milk; pour over layers. Bake at 350 degrees for 1 to 1½ hours.

Mrs. T. H. M., Foster Co., N.D.

POTATO CAKES

- 2 c mashed potatoes
- 1 egg, beaten
- ½ c cream style corn
- ½ tsp. salt
- ¼ tsp. baking powder

Mix together all ingredients. Drop by spoonfuls into melted butter and fry slowly until golden brown. If batter is too thin add more flour.

Mrs. R. H., Mercer Co., N. D.

POTATOES CHANTILLY

- 3½ lbs. potatoes, peeled
- 1 c hot milk
- 1 tbsp. butter
- few drops onion juice
- ½ tsp. white pepper
- ½ tsp. salt
- 1 c heavy cream
- 1 c grated pasteurized process cheese
- salt
- pepper
- paprika

Cook potatoes in boiling salted water. When done, mash with milk, butter, onion juice and seasonings. Beat cream and cheese together; season to taste. Pile cheese mixture lightly on mashed potatoes and bake at 450 degrees a few minutes to brown.

Ms. M. H., Lincoln Co., S. D.

Potatoes/**VEGETABLES**

POTATO CORN BAKE

4 c thinly sliced potatoes
1 (No. 303) can cream style corn
2 tbsps. flour
salt and pepper
1 c cream or rich milk

Arrange in alternate layers in buttered 1½-quart casserole the potatoes, corn and flour. Season. Pour cream or milk over top. Bake at 375 degrees for 1 hour.

Mrs. W. B., LaCrosse Co., Wis.

Sweet Potatoes/**VEGETABLES**

CANDIED SWEET POTATOES

5 medium-sized sweet potatoes
¼ c orange juice or water
¾ c brown sugar
1/3 c butter or margarine
½ tsp. salt

Scrub and cook unpeeled sweet potatoes in covered saucepan until done. Drain by holding each potato on a fork and peel off skin. Halve potatoes lengthwise and arrange in a greased casserole. Heat remaining ingredients to boiling; pour over potatoes. Bake at 375 degrees for 30 minutes

Mrs. E. H., Burke Co., N. D.

Squash/**VEGETABLES**

SQUASH PUDDING

3 tbsps. butter or margarine
3 tbsps. flour
½ tsp. salt
¼ tsp. pepper
¼ tsp. nutmeg
1 c milk
1 pkg. frozen squash, thawed
1 c soft bread crumbs
1 tsp. grated onions
3 egg yolks, slightly beaten
3 egg whites, stiffly beaten

Make a cream sauce of the butter or margarine, flour, seasonings and milk. Add squash, bread crumbs, onion and egg yolks. Fold into egg whites and pour into greased ring mold. Set in a pan of hot water and cover mold with a heavy paper or foil. Bake at 475 degrees for 1 hour, or until set. Remove from water and let stand 5 minutes. Turn out on plate. Fill center with small buttered beets.

Mrs. E. M. B., McPherson Co., S. D.

VEGETABLES/Squash

BAKED ACORN SQUASH

3 acorn squash
6 tbsps. syrup or honey
1 tsp. salt
1 tsp. sage
1 lb. pork sausage
bread crumbs

Wash and halve each squash. Remove strings and seeds. Put 1 tablespoon honey or syrup in each half and sprinkle with salt and sage. Fill cavity with pork sausage and top with bread crumbs. Place halves in baking pan and add about 1 inch of water. Cover and bake at 400 degrees for 40 minutes. Remove cover and allow to brown.

Mrs. S. N., Stutsman Co., N. D.

VEGETABLES/Rice

BAKED RICE

4 c water
¼ tsp. salt
1 c washed rice
½ c washed raisins
½ c sugar
1 egg
1 c sweet cream
1 tbsp. butter

Bring water and salt to a boil. Add rice and raisins. Cook for 10-15 minutes, stirring often. Drain. Mix together sugar, egg and cream. Add rice and raisins. Pour into buttered casserole. Top with butter. Bake at 350 degrees for 45 minutes to 1 hour.

Note: 1 teaspoon vinegar added to the water when boiling rice will make whiter rice.

Mrs. E. R., Morton Co., N. D.

VEGETABLES/Turnips

TURNIP SURPRISE

4 c boiled, mashed turnips
6 tbsps. butter
2 tbsps. sugar
2 tsps. salt
¼ tsp. pepper
2 c bread crumbs
4 eggs, well beaten

Combine ingredients in order given, reserving ½ cup bread crumbs. Mix well. Turn into greased 1½-quart baking dish; top with remaining crumbs. Bake at 400 degrees for 20 minutes.

Mrs. A. V., Pennington Co., S. D.

SHORT-CUT ZUCCHINI

1 lb. zucchini
¼ c butter or margarine
¼ tsp. garlic salt
⅛ tsp. pepper
2 tbsps. water
2 tbsps. grated Parmesan cheese

Slice zucchini (do not peel). Melt butter; add zucchini, garlic salt, pepper and water. Cover tightly and simmer over low heat about 10 minutes. Sprinkle with cheese. Simmer an additional 5 minutes.

Mrs. J. U., Grant Co., N. D.

BAKED ZUCCHINI

3 zucchini
salt
pepper
1 tbsp. butter
¼ c chopped onion
½ c chopped celery
1 c ground beef or leftover meat
1 slice bread, cubed
1 (8-oz.) can tomato sauce
¼ c grated Parmesan cheese

Cook zucchini in boiling salted water about 3 minutes. Do not peel. Cut in half lengthwise. Scoop out some of pulp; add salt and pepper. Melt butter; add onion, celery and meat. Brown lightly. Add bread, scooped out pulp and tomato sauce. Stuff mixture in zucchini. Place in shallow baking dish. Sprinkle with cheese. Bake at 375 degrees for 30 minutes, or until tender.

Mrs. J. U., Grant Co., N. D.

Miscellaneous/VEGETABLES

GARDEN VEGETABLES WITH BACON

2 c fresh green beans, cut in 1-inch pieces
1 c water
salt
2 c diced celery
2 c fresh corn, cut from cob
4 bacon slices
1 tsp. seasoned salt
1 tbsp. chopped parsley

Cook green beans in salted boiling water for 10 minutes. Add celery and corn; cook, covered, for 10 minutes. In small skillet, saute bacon until crisp. Remove bacon, reserving drippings. Drain and crumble bacon. Add drippings, bacon, salt, pepper and parsley to drained vegetables. Toss lightly.

Mrs. K. S., Cass Co., N. D.

LEMON-BUTTER VEGETABLE SAUCE

¼ c butter
4 tbsps. canned or frozen lemon juice
salt
pepper
2 tbsps. blanched, shredded almonds

Beat butter until soft. Add lemon juice slowly and continue beating. Add salt and pepper and almonds. Serve over individual portions of cooked broccoli, asparagus or other vegetables.

Mrs. M. K., Fall River Co., S. D.

MUSHROOM CHEESE SAUCE

2 tbsps. butter
1 c sliced, fresh mushrooms or 1 (4-oz.) can mushroom stems and pieces
2 tbsps. flour
1 c milk
½ tsp. salt
½ c shredded sharp pasteurized process cheese

Saute mushrooms in melted butter; stir in flour. Add milk, stirring constantly. Cook and stir until sauce is smooth and thickened. Add salt and cheese; stir until cheese is melted. (If canned mushrooms are used, drain and add to sauce.) Serve hot over cooked cauliflower or green beans.

Mrs. J. D., Tripp Co., S. D.

QUICK VEGETABLE SAUCE

1 can cream of mushroom soup
1 small can deviled ham
¼ c mayonnaise

Blend all ingredients. Warm and spoon over hot, cooked vegetables.

Mrs. M. P., Yankton Co., S. D.

TOMATO HOLLANDAISE

½ c mayonnaise
salt
pepper
2 tbsps. ketchup
½ tsp. grated onion
¼ c water

Combine mayonnaise, salt and pepper, onion and ketchup in top of double boiler. Gradually add water, stirring constantly. Cook 5 minutes. Be careful not to overcook. Serve on hot green beans, carrots, onions, cabbage or spinach.

Mrs. E. S., Corson Co., S. D.

SOUR CREAM DRESSING

1 c sour cream
1 tbsp. chopped chives or minced onion
2 tbsps. vinegar
½ tsp. salt
¼ tsp. black pepper
¼ tsp. sugar
½ c mayonnaise or salad dressing

Mix together all ingredients except mayonnaise carefully. Fold in mayonnaise. Use as a dressing for baked potatoes, sliced cucumbers, lettuce and seafoods of all kinds.

Ms. B. D., Hettinger Co., N. D.

PRESERVING

"Putting up" fruits and vegetables is a time-honored tradition among farmwives. The pungent odor of vinegar and heavy steam from the canner turn many country kitchens into one-woman factories for days each summer. But the hard work pays off in pantry shelves lined with jars of home canned foods for winter eating. The preserving recipes in this section are Dakotas' finest, perfected by years of practice.

Pickles
Relishes

Preserves
Jams & Jellies

BEET PICKLES

5 qts. cooked beets
3 c vinegar
4 c sugar

To cook beets, wash, rinse several times; place in porcelain kettle; cover with cold water and boil until tender. Cool; remove skin and stems. Make a syrup of vinegar, sugar and 2 cups of the water the beets were cooked in. When syrup starts to boil, add beets. Bring to boil again; pack in jars and seal.

Note: By using a porcelain container, beets do not turn dark in spots and liquid is transparent.

Miss A. H., Griggs Co., N. D.

BEST IN THE WORLD DILLS

cucumbers
dill
4 c vinegar
1 c sugar
½ c salt
1 c water

Soak cucumbers in ice or very cold water for 1 hour. Pack with dill in jars. Make a brine of vinegar, sugar, salt and water. Pour hot over pickles; seal in jars.

Miss A. H., Griggs Co, N. D.

CAULIFLOWER PICKLES

8 c cauliflower
4 c vinegar
2 c brown sugar
4 c sugar
1 stick cinnamon, broken
2 tbsps. whole cloves
½ c salt
½ tsp. turmeric
¼ c mixed spices
1 tsp. ground cinnamon
2 large sweet peppers
4 qts. coarsely sliced cucumbers
4 c sliced or whole small onions

Cook cauliflower in salted boiling water for 10 minutes. Combine remaining ingredients; boil 10 minutes. Pack in jars with cauliflower and process in water bath for 10 minutes.

Mrs. G. R., Grand Forks Co., N. D.

CITRON PICKLES

4 qts. cirton rinds, pared and sliced
2 tsps. salt
water

Syrup:
2 c sugar
1 c vinegar
mixed pickling spices (tied in bag)

Combine citron rinds, salt and enough water to cover. Boil until tender; drain well. Combine syrup ingredients; cook slowly for 15 minutes. Put rinds into hot jars; cover with hot syrup. Seal.

Mrs. E. H., Griggs Co., N. D.

CRISP LIME PICKLES

cucumbers, sliced
lime
water
Syrup:
4½ lbs. sugar
8 c vinegar
1 tbsp. salt
1 tsp. celery seed
1 tsp. whole cloves
1 tbsp. mixed pickling spices

Cover cucumbers in solution of 1 cup lime to 1 gallon water. Let soak 24 hours; rinse well. Cover with fresh water; let stand 3 hours. Drain. Combine syrup ingredients; heat and pour over cucumbers. Let stand overnight. Boil in syrup about 40 minutes or until pickles begin to look clear. Pack in sterile jars and seal.

Mrs. M. W. C., Aurora Co., S. D.

CRISP SWEET PICKLES

cucumbers
water
1 c vinegar
2 c sugar
¼ tsp. turmeric powder
1 tsp. mixed spices
1 tsp celery seed

First morning: pour boiling water over cucumbers. Let stand overnight. Drain and pour on fresh boiling water; repeat for 6 mornings. On the seventh morning add enough salt so the water does not taste flat. On the eighth morning, cut into smaller pieces and pack in jars. Drain upside down for awhile to let excess water off. Heat vinegar and sugar; add remaining ingredients. Pour hot over cucumbers; seal.

Miss A. H., Griggs Co., N. D.

TRANSPARENT PICKLES

6 doz. cucumbers, thinly sliced
water
salt
alum
Syrup:
6 c water
6 c white vinegar
10 c sugar
2 tsps. whole cloves
2 tsps. whole allspice
2 sticks cinnamon (tied in bag)
2 tsps. salt
2 tsps. celery seed

Soak cucumbers in 1 gallon cold water and 1 cup salt for 7 days, stirring once a day. Wash; put in water to cover, using 1 gallon water to 1 tablespoon alum. Boil 10 minutes; drain. Wash cucumbers. Combine syrup ingredients; pour over cucumbers. Boil until transparent. Put into jars and seal while hot.

Mrs. A. G., McHenry Co., N. D.

GREEN OR YELLOW PICKLED BEANS

beans
1 c vinegar
1 c sugar
1 stick cinnamon
mixed pickling spices

Boil beans. Combine 1 cup juice from beans, vinegar and sugar; boil. Add beans. Bring to boil again; add cinnamon. Cool and seal in jars.

Mrs. R. M., Hanson Co., S. D.

MIXED VEGETABLE PICKLES

water
salt
4 c sliced cucumbers
4 c chopped onions
4 c chopped green tomatoes
1 cauliflower
4 c sliced celery
4 c cooked and sliced carrots
4 c cooked string beans
1 can kidney beans
4 c sugar
4 c vinegar
1 tsp. mustard seed
1 tsp. cinnamon buds or 1 stick cinnamon

Soak overnight in a brine (made of 1 cup salt to 1 gallon water), the cucumbers, onions, green tomatoes and cauliflower. Drain; add celery, carrots, string beans and kidney beans. Heat remaining ingredients. Add vegetable mixture; simmer 10 minutes. Put into jars and seal.

Miss A. H., Griggs Co., N. D.

TURMERIC PICKLES

12 cucumbers, sliced
1 tsp. celery seed
1 tsp. mustard
½ tsp. pepper
2 c vinegar
1 tsp. salt
1 tsp. ginger
½ tsp. turmeric
1 c sugar
6 onions, sliced

Let cucumbers stand in salt brine for 1 hour. Boil together remaining ingredients, except onions. Add onions and cucumbers. Boil a few minutes more. Put into jars and seal.

Mrs. F. W., McIntosh Co., N. D.

MOCK OLIVES

4 c large green plums
1 tbsp. salt
1 c cold vinegar

Pack fruit into 1-quart jar. Mix together salt and vinegar; pour over plums. Add water to fill jar; seal. Let stand at least 2 weeks before using.

Ms. A. J. J., Rolette Co., N. D.

MUSTARD PICKLES

- 2 qts. peeled, seeded and cubed large cucumbers
- salt
- 1 qt. peeled and diced carrots
- 1 qt. diced onion
- 1 qt. cut-up string beans
- 1 qt. peas

Dressing:
- 1 qt. vinegar
- 3 1/3 c sugar
- 2 tsps. celery seed
- 3 tbsps. mustard seed
- 1 tbsp. turmeric
- ½ c flour

Sprinkle cucumbers with salt; let stand several hours. Cook carrots until tender in salted water. Drain cucumbers and carrots; add remaining vegetables.

Dressing: Combine all dressing ingredients; boil. Add vegetables; heat through. Seal in hot, clean jars.

Mrs. D. F., Cass Co., N. D.

WATERMELON PICKLES

- 1 whole watermelon or watermelon rind, peeled and sliced
- dill
- red pepper
- garlic
- 2 tsps. sugar
- 1 tsp. salt
- ¼ c vinegar
- 2 c water

Pack watermelon slices into jars with dill, red pepper and garlic. Mix together remaining ingredients; boil. Pour hot over watermelon; seal. Let jars set in hot water for 30 minutes, do not boil.

Mrs. C. G., Bon Homme Co., S. D.

ZUCCHINI PICKLES

- 4 qts. zucchini
- 6 large onions, sliced
- 2 green peppers, cubed
- 1 red pepper, chopped or sliced
- 2 cloves garlic, finely chopped
- salt
- 5 c sugar
- 1½ tsps. turmeric
- 2 tsps. mustard seed
- 3 c white vinegar

Combine vegetables and garlic. Salt well and let stand in ice water for 3 hours; drain. Combine remaining ingredients. Bring to boil and add vegetables. Cook a few minutes; seal in sterilized jars.

Mrs. D. K., Hutchinson Co., S. D.

HOT DOG OR HAMBURGER RELISH

15 cucumbers, peeled
3 onions
1 stalk celery
3 green peppers
4 carrots
½ c salt
2 c vinegar
3½ c sugar
1 tsp. celery seed
1 tsp. mustard seed
¼ tsp. turmeric

Grind together all vegetables; put into glass or stainless steel kettle and sprinkle with salt. Let stand overnight, uncovered. Drain; add remaining ingredients; bring to boil. Pour into pint jars and seal immediately.

Mrs. G. F., Minnehaha Co., S. D.

CORN RELISH

corn from 18 ears sweet corn
2 large green peppers, seeded and chopped
2 sweet red peppers, seeded and chopped
4 onions, chopped
1 small head cabbage, chopped
1 c chopped celery
4 c vinegar
2 c brown sugar
2 tbsps. salt
3 tbsps. mustard

Mix together all vegetables; add remaining ingredients. Cook for 20-30 minutes, or until corn is tender, stirring occasionally. Pack into hot jars and seal.

Ms. E. H., Griggs Co., N. D.

CUCUMBER RELISH

12 large cucumbers
10 large onions
¼ c salt
2 c vinegar
2 c sugar
1 tsp. mixed spices
1 tsp. whole cloves
1 tsp. turmeric
1 tsp. celery seed (tied in bag)

Grind together cucumbers and onion. Add salt; let stand overnight; drain. Add remaining ingredients. Boil for 30 minutes. Put into jars and seal.

Mrs. A. G., McHenry Co., N. D.

BEET RELISH

4 c finely chopped beets
1 small cabbage, chopped
1 c ground red pepper, or
1 c grated horseradish
2 c sugar
2 tbsps. salt
2 tsps. mustard
2 c cider vinegar

Mix together all ingredients. Let stand 24 hours. Bring to a boil. Put in jars and seal.

Miss A. H., Griggs Co., N. D.

CARROT PRESERVES

4 c cooked carrots
2 oranges, with peels
3 c sugar
1 c vinegar

Grind together cooked carrots and oranges. Add sugar and vinegar. Boil for 30 minutes; seal in pint jars.

Mrs. R. W., McPherson Co., S. D.

ROSY RHUBARB PRESERVES

5 c red rhubarb, cut-up
1 c crush pineapple
3 c sugar
1 pkg. strawberry gelatin

Combine rhubarb, pineapple and sugar. Cook about 15 minutes, or until thick and clear. Remove from heat; stir in strawberry gelatin. Seal hot.

Mrs. M. H., Emmons Co., N. D.

RHUBARB-RASPBERRY JAM

5 c cut-up rhubarb
4 c sugar
1 pkg. frozen raspberries
1 pkg. raspberry jello

Combine rhubarb and sugar; let stand until dissolved. Add raspberries; boil for 5-8 minutes, stirring often. Remove from heat; add jello. Stir until jello dissolves. Pour into jars and seal.

Mrs. R. F., Clay Co., S. D.

RHUBARB AND ORANGE CONSERVE

2 lbs. rhubarb, cut in ½-in. cubes
3½ c sugar
grated rind and juice of 2 oranges
grated rind and juice of 1 lemon
½ c nut meats

Combine rhubarb, sugar, orange and lemon rind and juice. Slowly cook until mixture is clear and thick. Add nut meats; cook 5 minutes longer. Pour into sterilized glasses; cover with paraffin.

Mrs. R. O., Benson Co., N. D.

RHUBARB-PINEAPPLE MARMALADE

4½ c rhubarb
3 c sugar
1 tsp. grated lemon rind
2 tbsps. lemon juice
1½ c crushed pineapple

Combine all ingredients; cook slowly for 25 minutes or until thick and clear, stirring constantly. Pour into hot sterilized jars and seal.

Mrs. I. O. D., Norman Co., Minn.

SUNSHINE STRAWBERRY PRESERVES

2 tbsps. lemon juice
4 c hulled strawberries
4 c sugar

Combine all ingredients; heat slowly to boil. Cool rapidly, stirring constantly for 8-10 minutes. Pour berry mixture into platters or shallow containers. Cover with glass or cellophane wrap, allowing a small space for evaporation. Set preserves in sunshine, stirring occasionally during the day; bring in at night. Repeat for a total of 3 days. Pack into sterilized jars and seal.

Mrs. C. W., Renville Co., N. D.

GRAPE JELLY

2 c grape juice
3 c sugar

Boil grape juice and sugar for 5 minutes. Remove from heat; stir for 15 minutes. Pour into jars and seal with paraffin.

Mrs. M. D., McIntosh Co., N. D.

TOMATO JAM

5 lbs. ripe tomatoes, peeled and diced
5 c brown sugar
2½ c vinegar
1 tbsp. cloves
1 tbsp. allspice
stick cinnamon (tied in cheesecloth)
3 c chopped seedless raisins

Combine all ingredients except raisins. Boil slowly for 2 hours. Add raisins; boil 1 hour longer. Remove spices; pour into sterilized jars and seal hot.

Mrs. D. M., Logan Co., N. D.

GREEN TOMATO PRESERVES

1 lb. green tomatoes
1 lb. sugar
2 tbsps. preserved ginger
2 lemons, thinly sliced

Pour boiling water over tomatoes; skin and quarter. Add sugar and let stand overnight. Pour off syrup. Boil syrup again until thick; skim. Add tomatoes and ginger. Cut into small pieces; add lemon slices. Boil until tomatoes are clear. Bottle and seal while hot.

Mrs. H. C., Sully Co., S. D.

PRESERVING/Jams & Jellies

APPLE BUTTER

Peel, core and cut up apples. Put small amount of water in bottom of kettle. Cook apples until mushy. Measure 6 cups sugar and 6 cups apple pulp. Add cinnamon candy to suit taste. Cook until thick; seal in jars.

Mrs. F. J., Brookings Co., S. D.

PRESERVING/Miscellaneous

TOMATO JUICE

- 15 pts. juice
- 7 tbsps. sugar
- 5 tbsps. vinegar
- 3 tbsps. salt

Combine all ingredients; mix well. Pour into 7 1-quart jars and process for 20 minutes in open kettle or pressure cook at 5 pounds for 10 minutes.

Mrs. L. H., Wadena Co., Minn.

KETCHUP

- tomatoes
- ½ c vinegar
- ½ c sugar
- 1 tsp. salt
- ½ tsp. ginger
- ½ tsp. nutmeg

Boil ripe tomatoes in own juice until soft. Strain through flour sieve to make 1 quart juice. Add remaining ingredients; cook until thick. Seal in sterilized jars when cold.

Mrs. H. B., Minnehaha Co., S. D.

HOME CANNED PORK AND BEANS

- 4 lbs. dry beans
- 2 qts. tomato juice
- 2 tbsps. molasses
- 4 c water
- 1 large onion, chopped
- 1 c sugar
- red food coloring
- 1 tbsp. coarse salt
- thick slices bacon

Soak beans overnight in cold water. Boil for 45 minutes. Cook together and strain tomato juice, molasses, water, onion and sugar. Drain beans. Add food coloring. Pour sauce over beans. Add salt. Fill jars ¾ full; add 2 slices bacon to each jar. Pressure cook at 10 pounds for 40 minutes, or cold pack 1½ hours.

Mrs. I. N., Emmons Co., N. D.

Make It From Scratch

"You can't beat homemade." This is the philosophy that has earned farmwives their reputation as the best cooks in the world. Of necessity our pioneer grandmothers started from scratch in preparing all the food the family consumed. However, many homemakers today follow this practice because they are convinced everything tastes better that way. In this section you will find recipes for homemade pops and wines, cheese, noodles and even soda crackers.

Dairy **Wine and Beverages** **Miscellaneous**

HOMEMADE CHEESE

Squeeze moisture from cottage cheese that has been made from 2½ gallons milk. Crumble the cheese. Add 1½ teaspoons soda. Mix well. Let set 1 hour. Add ¼ cup butter. Put into heavy skillet (cast iron is best) on low heat until cheese and butter are melted. Add 1½ cups heavy cream; salt to taste. Leave on heat until all is well blended. Add a few pimientoes and butter coloring.

Mrs. T. H., Perkins Co., S. D.

COTTAGE CHEESE

Sour milk in container. Put container over very low heat; stir slightly until warm. Let cool in container; pour into sieve. Let stand 1 hour until juice is out. Mix with cream, salt and pepper.

Mrs. F. E., Traill Co., N. D.

SOUR CREAM FROM DRY MILK

- 2/3 c dry milk
- ¾ c water
- 1 tsp. vinegar

Beat ingredients well. Put in container; chill.

Mrs. N. A., Walworth Co., S. D.

WHIPPED BUTTER

Let ¼ pound butter stand at room temperature for 1 hour. Beat at low speed in electric mixer until large chunks smooth out. Gradually increase mixer speed to fastest position; whip until fluffy (about 8 minutes). Cover until ready to use. Makes ¾ cup.

Mustard Butter: Blend ¼ cup prepared mustard into whipped butter. Good on ham or cheese sandwiches.

Herb Butter: Blend 1 tablespoon minced parsley, 1 teaspoon lemon juice, ½ teaspoon savory and ¼ teaspoon salt into whipped butter. Good on hot rolls or potatoes.

Cinnamon Butter: Gradually beat 1 cup brown sugar and 2 teaspoons cinnamon into whipped butter. Good on pancakes or waffles.

Garlic Butter: Blend 1 clove garlic, minced, into whipped butter. Good on garlic bread or roast beef.

Mrs. E. P., Brookings Co., S. D.

MOCK WHIPPED CREAM

½ c margarine
½ c shortening
1 c milk, scalded and cooled
1 c sugar
1 tsp. vanilla

Beat margarine and shortening until light. Slowly add cooled milk; add sugar and vanilla. Beat well. May be frozen.

Mrs. P. R., Beadle Co., S. D.

NON-FAT WHIPPED CREAM

6 tbsps. powdered non-fat milk
1 c water
2 tsps. gelatin
1½ tbsps. cold water
¼ c sugar
1 tsp. vanilla

Dissolve milk in water; scald. Let gelatin soak in cold water; combine scalded milk, dissolved gelatin and sugar. Stir; chill in refrigerator until it jells. Beat mixture until it is the consistency of whipped cream. Add vanilla; whip again.

Mrs. B. H., Marshall Co., Minn.

CULTURED SOUR CREAM

1 pt. light cream
2-3 tbsps. fresh cultured buttermilk
1/16 tsp. salt

Sterilize all cups, spoons and utensils. Blend creams, buttermilk and salt thoroughly in sterilized equipment. Cover; let stand at room temperature until it thickens. Chill for future use; keep covered.

Fresh chives — Top potatoes with sour cream and chives.

Zippy Blue Cheese — Cream finely shredded Blue Cheese with sour cream. Good for baked potatoes or spinach.

Bacon — Crumble 3 or 4 strips of broiled bacon into 1 cup sour cream, add onion salt and Tabasco sauce to taste.

Horseradish — Mix ½ cup sour cream with 2 teaspoons horseradish, ¼ tsp. salt, ¼ teaspoon sugar, ½ teaspoon curry powder. Serve on vegetables or green salad.

Currant Jelly — Combine ½ cup sour cream with ¼ cup beaten currant jelly, 1 tablespoon lemon juice, ⅛ teaspoon salt. Use with fruit salad.

Mrs. J. K., Burleigh Co., N. D.

YOGURT

6 c water
2 c powdered milk
1 (5 1/3-oz.) can evaporated milk
3 tbsps. commercial yogurt

Combine all ingredients; mix well. Pour into jars; cover. Set in heavy pan of warm water coming to brim of jars. Bake at 105-120 degrees for 4 hours, or until texture of pudding.

Note: Last yogurt made or yogurt culture may be substituted for commercial yogurt.

Mrs. D.D., Bottineau Co., N. D.

Wines & Beverages/MAKE-IT-FROM-SCRATCH

DANDELION WINE

1 qt. flowers
1 qt. boiling water
3 qts. cold water
4 lbs. sugar
juice from 2 lemons

Pour boiling water over flowers; let stand overnight. Strain; add cold water, sugar and lemon juice. Store in stone jar for 6 weeks. After this it is ready to bottle.

Mrs. J. K., Nelson Co., N. D.

RHUBARB WINE

16 lbs. rhubarb
20 qts. water
16 lbs. sugar
6 lemons
6 oranges
raisins

Mash rhubarb; put into stone jar. Add water; let stand for 9 days. Strain; add remaining ingredients. Cover; stir once or twice a day. Let stand at least 6 weeks. Skim well; bottle.

Mrs. J. K., Nelson Co., N. D.

ORANGE POP

1 yeast cake
5 gals. warm water
5 lbs. sugar
juice from 6 lemons
1 bottle orange nectar

Dissolve yeast in warm water; add remaining ingredients. Let set 5 minutes. Pour in bottles and cap. Keep in warm place for 5 days.

Mrs. A. R., Burleigh Co., N. D.

ROOT BEER

1 cake yeast
½ c warm water
5 gals. water
8 c sugar
6 lemons
2 pkgs. root beer flavored powdered drink

Soak yeast in warm water. Stir yeast mixture into water. Add remaining ingredients; mix well.

Mrs. W. W., Mead Co., S. D.

MAKE-IT-FROM-SCRATCH/Miscellaneous

JELLO POPSICKLES

2 c hot water
1 pkg. cherry gelatin
1 pkg. cherry-flavored summer drink powder
½ c sugar
2 c cold water

Combine hot water, cherry gelatin, powder and sugar; add cold water. Pour into ice cube trays; freeze partially. Insert ice cream sticks; freeze firm.

Mrs. B. K., Deuel Co., S. D.

FUDGESICLES

3 c milk
½ c sugar
1 pkg. chocolate pudding
1 c whipped cream

Combine milk, sugar and pudding. Cook until thick. Cool. Add whipped cream. Pour into ice trays; freeze.

Mrs. L. B., Mercer Co., N. D.

HONEY GRAHAM CRACKERS

1 c shortening
1 c honey or sugar
2 eggs
½ tsp. baking powder
1 tsp. soda
¼ tsp. salt
1 c milk
5-6 c graham flour
white flour

Combine above ingredients except white flour; mix well. Roll dough out on board, using as much white flour as necessary. Bake at 425 degrees until done.

Mrs. E. S., Brown Co., S. D.

SODA CRACKERS

9 c flour
1 c lard
1 tsp. salt
1 tsp. soda
2 c cold water

Blend together flour and lard. Add remaining ingredients; mix well. Roll dough thin. Cut into squares; prick with fork. Bake at 425 degrees until done.

Mrs. E. S., Brown Co., S. D.

HOMEMADE MUSTARD

2 tbsps. flour
2½ tbsps. dry mustard
sugar to taste
salt to taste
vinegar to taste
horseradish (optional)

Brown flour, stirring constantly. Add remaining ingredients. Store in small jar.

Mrs. E. L., Shannon Co., S. D.

PREPARING SUNFLOWER SEEDS

sunflower seeds
2 qts. distilled water
¼-½ c salt

Cover raw seeds with distilled water and salt; simmer 2 hours. Drain seeds well on paper towels. Place on cookie sheet; roast at 300 degrees for 30-40 minutes. Stir occasionally.

Mrs. O. E. C., Cavalier Co., N. D.

NOODLES

2 egg yolks, slightly beaten
1 tsp. lard
¼ tsp. salt
2 tbsps. cream or milk
¼ tsp. baking powder
flour

Combine above ingredients, using enough flour to make stiff dough. Roll dough out on lightly floured dish towel until thin. Allow to dry until edges are hard; cut into noodles. Spread on towel; dry. Cook or store.

Miss D. B., Stutsman Co., N. D.

SYRUP

1 c brown sugar
1 c sugar
1 c water
½ tsp. vinegar

Combine sugars and water. Boil to 225 degrees. Remove from heat; add vinegar to prevent sugar crystals.

Mrs. H. E. E., Cass Co., N. D.

Index